T0228022

Sciences from Below

NEXT WAVE: NEW DIRECTIONS IN WOMEN'S STUDIES
A series edited by Inderpal Grewal, Caren Kaplan,
and Robyn Wiegman

Sandra Harding

SCIENCES FROM BELOW

Feminisms, Postcolonialities, and Modernities

Duke University Press   Durham and London   2008

© 2008 Duke University Press
All rights reserved
Printed in the United States of America on acid-free paper ∞
Designed by C. H. Westmoreland
Typeset in Scala and Scala Sans by Keystone Typesetting, Inc.
Library of Congress Cataloging-in-Publication data appear
on the last printed page of this book.

CONTENTS

— — — —

Acknowledgments   vii

Introduction: Why Focus on Modernity?   1

I.   PROBLEMS WITH MODERNITY'S SCIENCE AND POLITICS
     Perspectives from Northern Science Studies
     1. Modernity's Misleading Dream: *Latour*   23
     2. The Incomplete First Modernity of Industrial Society: *Beck*   49
     3. Co-evolving Science and Society: *Gibbons, Nowotny, and Scott*   75

II. VIEWS FROM (WESTERN) MODERNITY'S PERIPHERIES
     4. Women as Subjects of History and Knowledge   101
     5. Postcolonial Science and Technology Studies: *Are There Multiple Sciences?*   130
     6. Women on Modernity's Horizons: *Feminist Postcolonial Science and Technology Studies*   155

III. INTERROGATING TRADITION: CHALLENGES AND POSSIBILITIES
     7. Multiple Modernities: *Postcolonial Standpoints*   173
     8. Haunted Modernities, Gendered Traditions   191
     9. Moving On: *A Methodological Provocation*   214

Notes   235
Bibliography   257
Index   281

ACKNOWLEDGMENTS

— — — —

THIS BOOK COULD NOT HAVE BEEN WRITTEN without the assistance, support, and critical thought of many scholars, activists, and friends working in the fields of science studies, feminisms, postcolonial studies, and modernity studies, especially those working at the intersections and shared paths of these fields. Particularly valuable were the reviews of all or parts of the penultimate manuscript provided by David Hess, Françoise Lionnet, Sara Melzer, Clayton Pierce, Hilary Rose, E. Ann Tickner, and Sharon Traweek. Doug Kellner and Gail Kligman generously shared materials and critical feedback with me during the years I was working up this project.

Lecture audiences at many conferences and universities helped me to see the importance of the issues here and how to position them in useful ways. To two I owe a special debt: the "Modernity in Transit" conference at the University of Ottawa in 2000, organized by Pascal Gin and Walter Moser, which enabled me to think of launching this project in the first place; and the commentators and audience at the Faculty Seminar Series of the UCLA Center for the Study of Women, organized by Kathleen McHugh in April 2007, who helped me improve the last two chapters.

Tanner LeBaron Wallace and Tara Watford helped me track down research materials. Nancy Lee Sayre gave the manuscript a fine copyediting.

My lovely housemates, Emily and Eva Harding-Morick, nourished my soul on a daily basis.

INTRODUCTION

Why Focus on Modernity?

— — — —

THIS IS A BOOK ABOUT WESTERN MODERNITY and the ways it remains haunted by anxieties about the feminine and the primitive, both
of which are associated with the traditional. Northern philosophies of
science and technology have been complicitous in establishing and
maintaining these haunting specters. Scientific rationality and technical expertise are presented in these philosophies as the one-way time
machines that supposedly enable elite Westerners and men around the
globe to escape the bonds of tradition, leaving behind for others the
responsibility for the flourishing of women, children and other kin,
households, and communities, and for the environments upon which
their flourishing depends. These others must do the kind of reproductive and "craft" labor necessary to raise acceptably human children of a
particular culture, maintain community social bonds, and "suture" the
new—such as railroads or electric cars—to the familiar conceptually,
materially, morally, and politically. These others are mostly women and
non-Western men. How can Western modernity hope to deliver social progress to women and non-Western men when its most valued
achievements are measured in terms of its distance from the interests,
needs, and desires of the very humans who produce and reproduce
human life and the world around us in ways that make Western modernity possible?

This is not to say that Westerners or men in non-Western cultures individually hold such views. Many have struggled, often at great costs, to enable their women, children, kin, households, communities, and local environments to flourish. Rather, the point here is that the institutions of Western modernity and their scientific and political philosophies, designed by and for men in elite classes, persistently create meanings and practices of modernity which create fearful specters of "the feminine" and "the primitive." Even purportedly progressive scientific and technological projects, in the North and the South, are doomed to fail when they do not critically engage with the specters of modernity.[1]

1. MODERNITY BEYOND POSTMODERNITY

The topic of modernity can seem unpromising for a number of reasons at this moment, however. For one thing, it has already been a constant topic of discussion for many decades, ever since postmodernism became a theoretical project able to bring into sharp focus widespread discontents with modernity. Feminisms and postcolonialisms seem to share with postmodernism skepticism about modernity's idealized rational man; his propensity for grand narratives that presume to provide a universally valid official history and to be able to predict the future from a supposedly culture-free perspective; his assumptions about an innocent core self which exists prior to its encounter with culture; and the various ontological, epistemological, political, and ethical theories and practices which flow from this familiar discourse (Flax, *Thinking Fragments*). So there are good reasons why feminisms and postcolonialisms are frequently labeled postmodern. Yet in other respects these social movements also seem firmly lodged in modernity, or at least unwilling to commit themselves to the side of either modernity or postmodernity. They seem unwilling to engage in the luxury of postmodernist disillusion with politics and its silence in the face of needed social justice projects. Could there be anything further useful to be said about this already widely discussed dispute?

Another discouraging problem is that terms such as *modernity, the modern, modernization,* and *modernism* seem to be used in different ways by speakers in different disciplines, different political orientations, and even different languages (Friedman, "Definitional Excursions"). It is hard to know just what is meant by embracing or rejecting modernity, modernization, or modernism in the face of the confusing references and meanings these terms have. Here we will shortly clear the ground for the discussions of modernity to follow by identifying some basic characteristics of the phenomena to which these terms refer. But before we do so, let us note five reasons to pursue discussions of modernity beyond the point where postmodernism ended.

First, while modernity was forced to turn and face postmodernism by the latter intellectual movement, its earlier and ongoing contrast with tradition and the pre-modern was largely obscured and, at any rate, not interrogated in the postmodernity discussions. Yet the modernity vs. tradition binary remains powerful today in shaping research in the natural and social sciences and their philosophies as well as in the public policy which such research serves. Such work typically treats the needs and desires of women and of traditional cultures as irrational, incomprehensible, and irrelevant—or even a powerful obstacle —to ideals and strategies for social progress. No wonder modernity's social progress has been delivered to only such a small minority of the world's citizens.

Second, this binary needs examination because modern discourses are haunted by specters of the feminine and the primitive. Objectivity, rationality, good method, real science, social progress, civilization—the excellence of these and other self-proclaimed modern achievements are all measured in terms of their distance from whatever is associated with the feminine and the primitive. Western sciences and politics, and their philosophies, need an exorcism if they are to contribute at all to social progress for the vast majority of the globe's citizens! Here such an exorcism is performed through critical examinations of tradition. Such a project has become possible only with the emerging insight that modernization is not identical to Westernization. This project abandons the narratives of exceptionalism and triumphalism which have been favored in the West. Western modernity is not the only modernity which has emerged around the globe and which has admirable features. And Western modernity has brought not only great benefits to some, but also great disasters to many. To understand modernity more fully, it turns out that we have to focus on tradition.

What are exceptionalism and triumphalism? By exceptionalism I mean the belief that Western sciences alone among all human knowledge systems are capable of grasping reality in its own terms—"cutting nature at its joints," as philosophers of science typically enjoy referring to the matter. According to this view, only modern Western sciences have demonstrated that they have the resources to escape the universal human tendency to project onto nature cultural assumptions, fears, and desires. Indeed, these research projects alone of all human inquiries into natural and social orders are entitled to be called sciences, according to the defenders of exceptionalism. Critics document just how such exceptionalists conflate Science with science. That is, the exceptionalists conflate the West's idealized understandings of its own practices with the universal human impulse to understand ourselves and the world around us in ways that permit effective interactions with such worlds. In contrast, the critics argue that "all people operate within the domains of magic, science, and religion" (Malinowski, *Magic* 196; quoted by Nader, *Naked Science* 5). Modern Western sciences are just one set of sciences today, albeit powerful ones, among the many others that have existed and do today around the globe. Moreover they are not constituted entirely by Europeans or within European civilizations; in fact they owe great debts, mostly unacknowledged, to the science traditions that preceded them, especially those in Asia (see e.g., Hobson, *Eastern Origins of Western Civilisation*; Selin, *Encyclopedia of the History of Science*).

By triumphalism I mean the assumption that the history of science (which, for triumphalists, is thus the exceptionalist history of Western science) consists of a narrative of achievements. For triumphalists, this history has no significant downsides. From this perspective, Hiroshima, environmental destruction, the alienation of labor, escalating global militarism, the increasing gap between the "haves" and the "have nots," gender, race, and class inequalities—these and other undesirable social phenomena are all entirely consequences of social and political projects. The history of Western science proper makes no contribution to such social events and processes. These are a matter of the political and social uses of the pure knowledge which scientific inquiry produces. They are appropriately discussed under the heading of the applications and technologies of science, but not of sciences' representations of the natural world or distinctive (they say) methods of intervening in it. Exceptionalist and triumphalist assumptions about Western sciences are obviously mutually supporting. However, neither can today gather the support either in the West or elsewhere that they once could claim. Widespread scepticism about such histories and philosophies of modern sciences have prepared the ground for the issues about modernities and sciences I will raise here.

Even work that is otherwise innovatively progressive—work which understands, for example, that we must transform politics and social relations in order to transform sciences into more competent knowledge production and service to democratic social tendencies—remains captive to exceptionalism and triumphalism insofar as it distances itself from the insights of feminist and postcolonial science studies. Consequently, even this progressive work is doomed to failure since it does not access the resources necessary to bring about the projects of democratic political and scientific transformation to which these authors aspire. It provides analyses of modernity, its strengths and limitations, only "from above" when it avoids taking the standpoint of women and the world's other least-advantaged citizens on such topics. It is doomed to the loss of both competence and legitimacy in the eyes of the vast majority of the world's citizens—losses already fully under way today, as we shall see. The account here is intended to contribute to the different project of looking at modernity and its sciences "from below."

Returning to our list of reasons to pursue issues about modernity further, once exceptionalist and triumphalist narratives of Western history no longer can gather either empirical support or moral/political approval, Westerners must develop new notions of expertise, authority, and desirable speech which do not depend upon such narratives. Western ways of understanding the world are not always right or the best ways, and certainly not uniquely so. Westerners must learn how to make ourselves fit, and to be perceived to be fit, to enter into the democratic, pluricentric global dialogues from which global futures will emerge. This is a third reason to continue this project.

Fourth, we can come to see how modernity's Others have produced resources valuable for everyone who is interested in thinking about how to transform the modern social institutions we have into ones more suitable for today's and tomorrow's progressive global social relations. Feminist and postcolonial science and technology movements, separately and conjoined, will be the focus of such discussions here. What we can know about nature and social relations depends upon how we live in our natural/social worlds. And peoples at the peripheries of modernity—women and other marginalized groups in the West and peoples from other cultures—have lived differently, with distinctive kinds of interactions with the world around them, than those at the centers. This is not to say that the centers are all bad and the peripheries are all good. It is not to say that Western modernity has to date produced no still-desirable features, or that everything in traditional cultures is valuable. It is certainly not to say that Westerners should abandon the West and yet again seek salvation and innocence in the worlds of modernity's Others. Rather we need realistic reassessments of both Western and non-Western knowledge systems and the social worlds with which they are constituted rather than romantic evaluations of one and demonizations of the other. The point here is that "tradition" requires more realistic assessments than it has received within the horizons of Western modernities, and that such assessments by groups on those horizons already reveal rich resources for living together on this planet which have been ignored or disvalued in the West's modernity.

Finally, pursuing issues about modernity further in these ways raises new questions about postmodern discourses. They, too, will need to be reevaluated from the kinds of perspectives of the Others of Western modernity engaged here. To put my point another way, I am asking the field of science and technology studies to become even more controversial than it already is. Let us briefly recollect the sources of its existing controversiality.

2. A CONTROVERSIAL WORLD

This field has been controversial from its beginnings. Almost half a century ago it set out to show "the integrity" of high points in the history of modern science with their particular historical eras, as Thomas S. Kuhn (*Structure of Scientific Revolutions*) famously put the point. The new sociologies, histories, and ethnographies of science have revealed how scientific inquiry has been a social institution with many features of other social institutions (cf. Biagioli, *Science Studies Reader*). Subsequently, social constructivist tendencies in technology studies have shown how technologies are not merely value-neutral chunks of hardware; "artefacts have politics," as Langdon Winner ("Do Artefacts Have Politics?") argued (cf. MacKenzie and Wajcman, *Social Shaping of Technology*). The recent Science Wars provide one kind of testimony to how unsettling it can still be for many people—scientists and nonscientists alike, and whether they think of their political commitments as on the left, right, or center—to be asked to recognize that the highest achievements of the North's natural sciences are deeply permeated by distinctively historical social projects and practices.[2]

Feminist science studies has frequently been a target of such fears and criticisms. To be sure, any charges of continuing male supremacy are unnerving today to the many men and women who hope and believe that traditional patterns of discrimination in every field and profession mostly have ended. Any continuing signs of such discrimination are merely residues of those older patterns, they assume, and such residues are destined soon to disappear. Yet the challenges to the natural sciences remain especially troubling. Feminists have criticized the incompetence of the very standards of objectivity to identify widespread patterns of gender biases in the sciences. This charge strikes at the heart of what is generally considered to be most admirable about scientific research and its rationality—its methods of research. Thus they have also criticized the inadequacies of its standards for rationality, good method, and "real science." These criticisms focus not on the prejudices of individuals (unpleasant as those can be for their targets), but rather on the assumptions, practices, and cultures of institutions, and on prevailing philosophies of science. Of course it is scientific rationality and its standards for objectivity that also structure and set standards for the modern social institutions, principles, and practices that are regarded as most progressive in the modern industrialized societies of the North. If the objectivity and rationality of the natural sciences are questionable, so too is the progressiveness of the social institutions of which citizens of industrialized societies are most proud. For whom do modernity's sciences provide social progress?

Yet science and technology studies could usefully become much more controversial, and that is the recommendation here! There are two areas of appropriate analysis which have been under-addressed. This book sets out to explore how they can be used to turn familiar science studies and feminist frameworks into even more widely controversial topics of public discussion and debate. Such public discussion and debate are a necessity in societies aspiring to democracy and social justice, and in which the proposals for new destinies for the sciences and for women are the sites of both powerful yearnings and fearful anxieties.

One such understudied topic is the effect of Northern scientific and technological inquiry on peoples and cultures at the peripheries of Northern modernity. The experiences of these peoples have been articulated for several decades now through postcolonial science and technology studies and its feminist components. Though the concerns of this field can appear exotic and tangential to many people who are concerned with social equality, such an appearance is deceiving. These peripheries are more and more loudly "talking back" to the centers about such matters for both political and epistemological reasons. Fortunately, such topics are finally beginning to appear in mainstream science and technology studies conferences and publications.[3]

The other neglected topic in Northern science and technology studies is the modernity/tradition binary. A few scholars, on whom we will focus, have taken up the issue of whether and how (Northern) sciences and technologies are modern. Yet they, like the rest of the field, have largely stayed within the conceptual framework of modernity when it comes to issues about tradition, the premodern, and their conventional association with nature, the past, women, the feminine, the household, "the primitive," and loyalty to kin and "tribe." This binary creates horizons for Northern thought beyond which lie the irrational, the incomprehensible, and the unintelligible—namely, the worlds of the peoples neglected in the first topic as well as the worlds of women in the North. Here the interests of Northern women, on the one hand, and women and men in societies in the South, on the other hand, are conjoined (though they are of course not identical). The neglect of this second topic protects the neglect of the first. Hence the importance of the focus in this book on modernity and its Others, and the implications of such a discussion for the kinds of progressive transformations of Northern sciences and technologies which have been called for by so many groups committed to social justice.

To put the project of this book in other words, I want to "calibrate" to each other progressive tendencies in Northern science and technology studies, Southern science and technology studies, feminist work in both fields, and modernity studies.[4] I propose that each needs the success of the others for its own projects. What can each learn from the others?

3. MODERNITY: TEMPORAL OR SUBSTANTIVE?

Modernity and tradition will be defined and redefined again and again in the following chapters. Let us start off with some basic and competing ways in which modernity has been conceptualized. In the American Academy of Arts and Sciences issue on the topic of "multiple modernities" in its journal, *Daedalus*, Brian Wittrock identified problems with two ways of thinking about modernity which have long been widely influential.[5]

> When we speak of modernity and of modern societies, we seem to mean one of two things. First, we may speak as if we were giving an encompassing name to a whole epoch in world history, the modern age, as distinct from, say, the medieval age or classical antiquity. Such a terminology makes it legitimate to discuss questions as to when exactly the modern age may be said to have come into existence, what its origins may have been, or, indeed, if it has now come to an end. Second, we may speak as if we were actually characterizing distinct phenomena and processes in a given society at a given time. We may say that the technology used in some branch of industry of a country is modern but that patterns of family life are not. It is then an empirical question to determine to what extent different institutions and phenomena of a country may be described as modern. (Wittrock, "Modernity" 31)

Each of these notions is controversial. Such controversiality no doubt is one reason why many scholars prefer to avoid the whole topic of modernity and, rather, pursue their interests under other headings. But this strategy does not succeed in making the intellectual, social, and political power of the contrast between modernity and tradition go away—not even in their own work. Instead it comes to live a subterranean life, structuring thought, action, and public policy while remaining seemingly out of reach of public discussion and analysis.

Temporal modernity: Three referents

The temporal notion currently is used in the West with three distinct referents corresponding to the particular aspects of "the modern" which are in focus (cf. Friedman, "Definitional Excursions"). First, for philosophers and many historians of science, modern science begins in the seventeenth century with the scientific revolution of Copernicus, Galileo, Boyle, Harvey, and Newton, and modern philosophy begins with Hobbes and Descartes. The early modern philosophers engaged with implications of features of the world which new sciences such as astronomy and physics revealed, and they thought about some ways in which these new sciences did or could participate in the shifts in European social formations which they were experiencing. They thought about the new experimentalism in the sciences and about the new science movements of their day (Van den Daele, "Social Construction of Science"; Shapin and Schaffer, *Leviathan and the Air Pump*).

Yet some historians of science and technology would date the emergence of fully modern sciences later, in the bourgeois revolutions of the eighteenth century and the industrial revolution of the late eighteenth century and the early nineteenth. Copernicus, Galileo, Boyle, Harvey, and Newton are not yet truly scientists, they would hold. (Or, maybe there were two Western scientific revolutions?) These scholars are concerned especially with the new classes which supported the emergent democratic governments in the United States, France, and England; with urbanization; and with the increasing power of scientific technologies.

Modernization theorists, who produce the second kind of temporal notion of modernity, draw especially on this kind of history of modern sciences and technologies. Those concerned with modernizing traditional societies, for example in the Third World development policies of national and international agencies and institutions after the Second World War, always focus on transferring to underdeveloped societies (as they were characterized) Western scientific rationality and technical expertise in manufacturing, health care, agriculture, and other economic sectors. They take Western forms of modernization to be the only ones, as did their nineteenth-century forerunners such as Marx, Durkheim, and Weber. The nineteenth-century theorists created modern social sciences in their attempts to explain urbanization and industrialization. Consequently this conceptual legacy of contemporary social sciences seems to limit the critical resources that these sciences can bring to bear on modernity and modernization processes. For such theorists, as well as for some of their critics, modernization is identical to Westernization. Modernization means Western modernization, and "science" refers only to Western science. Like their nineteenth-century forerunners, the modernization theorists of the 1950s expected a gradual homogenization of global societies as Western forms of modernity disseminated around the globe. The term "modernization" has been used primarily to identify various pathways to change in "underdeveloped" societies, where it is associated with the transfer of Northern scientific rationality and technical expertise to the Southern societies.

By now, those Third World development policies grounded in modernization theory are widely criticized for further immiserating precisely the majority of the world's poorest citizens whom such policies were supposed to benefit (Amin, *Maldevelopment*; Sachs, *Development*

Dictionary; Escobar, *Encountering Development*). Feminist work has been an important part of this critique (Mies, *Patriarchy and Accumulation*; Shiva, *Staying Alive*; Sparr, *Mortgaging Women's Lives*; C. V. Scott, *Gender and Development*). Moreover, while modernity is now a global condition shaping how all societies engage with the world around them (Wittrock, "Modernity"), the expected homogenization of societies around the world has not occurred (Eisenstadt, "Multiple Modernities").[6]

Finally, for literary and cultural theorists, modernism refers to the late-nineteenth-century and early-twentieth-century movement which comes after romanticism. T. S. Eliot, James Joyce, Bauhaus architecture, Picasso, and Seurat are modernist. This literary and cultural movement has been the focus of what are perhaps the most developed analyses of the gender of modernity (e.g., Felski, *Gender of Modernity*; Jardine, *Gynesis*), though particular aspects of modernity have long been the topic of feminist sociologists, political theorists, and science theorists.

We must note that these different names for different pursuits of the modern seem to be characteristic in the English-speaking world. Yet in France "modernism" is used to refer to all three (Friedman, "Definitional Excursions"). "Postmodernism," also, can refer to any one of these three eras and its characteristic concerns. No wonder discussions of the modern among people from different disciplines can get confusing.

Substantive modernity
Deciding just when such temporal eras begin and end requires the specification of substantive criteria of the modern which some particular era does or does not meet. Thus the temporal notion collapses into or depends upon substantive criteria (Wittrock, "Modernity"). Substantive conceptions are controversial. Yet one can consistently find in the accounts of the post-World War II Western theorists and their nineteenth-century legacy a focus on the emergence of a differentiated social structure with political, economic, religious/moral, and educational (including scientific) institutions which are independent of family structures; the separation, therefore, of public and private spheres; and such democratic institutions as representative government, free elections, and a free press. Such conceptions also focus on a secular worldview, the idealization of universal instrumental rationality, and a social orientation toward the future rather than the past. They also

include several kinds of contradictory tendencies, such as the insistence on universal reason, yet also recognition and even toleration of the pluralism of rationalities, and a critical and self-critical attitude along with severe restrictions on the appropriate targets of such criticisms (Eisenstadt, "Multiple Modernities"; Wittrock, "Modernity"). Such contradictory tendencies are especially important for critics of modernity who would revise rather than turn their backs on the concept of modernity, as we shall see below.

Exceptions and complexities

Anyone who reflects for a moment about how the modernity vs. tradition binary structures issues in one's own area of expertise will immediately find exceptions and contrary tendencies which seem to refute the preceding attempt to organize the topic. For example, powerful cases have been made for the claim that no Western society has in fact fully achieved substantive modernity. (Here we have the ironic situation that modernity could not even fully come into existence before postmodernity declared its death.)[7] Moreover, modernization has proceeded unevenly around the globe, and even within Europe itself; parts of Spain and Russia were feudal monarchies well into the twentieth century. Furthermore, as we will see in Part II, there are good reasons to think that many of these substantive features of purportedly unique European modernity can in fact be found in non-Western societies, where they promiscuously mingle with local cultural features. Thus there seem to be many different modernities, always containing sciences and technologies, each with distinctive cultural features, "traditional" or not. Western modernities and sciences are just one among many possible organizations of post-premodern social realities and their sciences,[8] although they are today much more powerful in at least some respects than the alternatives. Thus, as the title of Wittrock's essay asks: "Modernity: One, None, or Many?" (31).

Yet the situation for modernities is even more complex. In Western societies today, modernity seems to be shrinking, not expanding as the classical theorists predicted. Many of the features taken to be required by the modernity paradigmatically found in Western societies seem to be disappearing. Modern institutions of the economy, politics, and education as well as science seem increasingly transgressive and simultaneously porous, as other such institutions come to permeate their practices and principles and they, in turn transgress in matters which were thought properly to belong to other institutions. We will see how this is so for the sciences, which increasingly appropriate political and economic functions and even religious institutional styles while permitting their own permeation by local, national, and international political and economic institutions. Is the West getting less and less modern?

Additional dimensions of discussions of the "modernity vs. tradition" binary will emerge in the chapters which follow. And we will propose some alternative conceptual frameworks that gain both cognitive and political power by avoiding the problems this binary poses as they better illuminate the empirical realities which appear abhorrent and/or unintelligible when seen only through the conventional binary.

Modernity is not the only central term in this study which has become a site of controversy. Three of the others are feminism, postcoloniality, and science. Actually, it would be better to speak of each in the plural in recognition of the diversity that is characteristic of each. Other controversial terms will be defined as we go along.[9]

4. MORE CONTROVERSIAL TERMS

Feminisms

Feminists share the belief that women, too, are fully human. This apparent platitude is in fact a revolutionary claim, the shock-value of which should not be underestimated. We will look at some of the consequences for sciences and modernities of actually taking women to be fully as human as their brothers. Feminists also share the belief that women's conditions in any particular historical context are largely a social matter. Of course there are biological differences between females and males in every species with two-sex reproduction. Yet women's conditions in societies are not, for the most part, caused by such biological differences (Fausto-Sterling, *Myths of Gender*). Feminists advocate for improving such social conditions. But beyond such general claims, there is considerable disagreement about just what women's conditions are, what the social causes of these conditions are, and how best to improve women's lives. Of course, there are similar controversies over the conditions of men's lives also; controversy over scientific and social explanations is not peculiar to feminists! In the West, one set of distinctive accounts of the nature, causes, and prescriptions for improvement of women's conditions can be found in the grand traditions of political theory. Thus Mary Wollstonecraft and John Stuart Mill developed Liberal (social contract) feminist theories in the context of the American and French Revolutions of the eighteenth century. Marx and Engels provided a powerful indictment of distinctive forms of women's class oppression, and socialist feminists created influential updates of these insights in the context of the New Left social movements and women's movements of the 1960s and 1970s. Radical feminists developed the first accounts of the at least partial autonomy of sexism and androcentrism from class oppression (Jaggar, *Feminist Politics*).

Beginning in the 1970s, a rich array of women-of-color, postcolonial, and transnational feminisms have emerged to map the distinctive effects of cultural difference and of class, race, and ethnic discrimination on women. They chart the many ways in which the lives of women in different cultures and classes around the world are linked through global networks of both empowerment and exploitation which advantage people of European descent, including women, and the elites in other cultures around the world. These feminisms offer illuminating explanations of how such processes occur and of the changes in conventional and much feminist social theory necessary to account adequately not only for the lives of women of color, but also for the lives of all the rest of the men and women in the world. Dominant groups cannot understand the nature and causes of their own social situations if they examine such topics only from their own "native" perspectives. It takes the standpoint of the oppressed and disempowered to reveal the objective natures and conditions of dominant groups. Modernity, its rationality, and its sciences look different from the standpoint of women's different social and cultural locations, and in the context of local and global systems of empowerment, oppression, and exploitation (Harding, *Feminist Standpoint Theory Reader*). Our paths through the modernity issues will seek out the theoretical and practical resources available by focusing on these differences.

It should be noted also that the term "feminism" is controversial in more than the ways obvious to women in the dominant groups in the West. Here male supremacists in the media (for example, on "talk radio") use it to conjure up radical, irrational man-haters who can only lead sensible young women into trouble. Yet in many parts of the world, feminism is seen as only a bourgeois Liberal movement inter-

ested in securing rights for rich women. The term is far too conservative for activists who care about the lot of poor and minority women. In
the United States, African American novelist Alice Walker thought it
important to introduce the term "womanist" to signal the possibility of
a social movement that was for women but designed to meet the needs
especially of African American women, whose concerns had not been
addressed in the prevailing "white" feminist movement. And Third
World male activists have frequently used the term "feminist" to dis-
parage what they see as foreign, Western, colonial, and imperial projects unsuitable for the culture-valuing and nation-building projects in
which they think women in their societies should involve themselves
(see chapters 6 and 8). Thus the term itself is a site for struggles over
political, economic, social, and cultural goals, practices, and resources.

Postcolonialities

European formal colonial rule ended for many societies around the
world only during the last half of the twentieth century. For some,
such as in Latin America, it ended in the early nineteenth century.
In Asia the history of imperialism and colonialism varies from country to country. (And still a number of other sub-cultures are agitat-
ing to escape rule by local dominant groups.) Is "postcolonial" the
most accurate way to designate these societies? Moreover, many would
say that "neocolonial" better describes Third World social systems in
that the interests and desires of Western nations, and especially the
United States, still dominate their economics, politics, and cultures. Of
course, all of those Western countries also have internal "colonies"
of disadvantaged groups, some only recent immigrants from former
formal colonies or from other economically disadvantaged societies.
"Postcolonial" can seem to all such less-advantaged groups to take a
position of an unwarranted triumphalism. "Decolonizing" or perhaps
even "postcolonializing" would be better terms to describe progressive
social movements and theoretical analyses (see Ashcroft, Griffiths, and
Tiffin, *Postcolonial Studies Reader*; Williams and Chrisman, *Colonial
Discourse*).

Yet I will use the term here for several reasons. For one, there is a
field of research and scholarship which calls itself "postcolonial studies." Discussions of modernity and of the sciences' roles in modernity
projects deserve to be part of this field. Moreover, the term clears a
discursive space for asking questions which have been otherwise diffi-

cult to raise. Postcoloniality can and must be a desire, a dream, and a vision long before it becomes a reality. (Think, for example, of descriptions of the United States or France as "democracies.") We shall see examples of such questions when we turn to postcolonial science and technology studies.

Sciences

I have already indicated that in contrast to the exceptionalist who thinks that this term must be reserved only for modern Western inquiries, I shall follow the lead of postcolonial science studies scholars who use it to refer to any and every culture's institutions and systematic empirical and theoretical practices of coming to understand how the world around us works.[10] Yet this expanded use of the term is controversial for still other reasons. The original producers of what has come to be called indigenous knowledge and traditional environmental knowledge do not refer to their activities as science. So one could regard the insistence here on doing so as another piece of Eurocentrism; if we are to take seriously the achievements of another culture, we have to talk about it in our terms, rather than theirs. Yet it can be valuable to do so in this case because such a practice levels the playing field by refusing to grant Western practices an entirely different, more highly valued, category of human inquiry. We can ask what we can learn about Western sciences and the inquiry practices of other cultures if we look at their similarities and their differences from a postcolonial standpoint instead of focusing, as Eurocentrists have done, only on their exceptional differences as Eurocentrists have identified them.

Of course the definition of what counts as science was not handed down from the heavens on stone tablets at the origins of modernity in the West. The early modern scientists called their work "natural philosophy." The term "scientist" only came into use in the early nineteenth century (Nader, *Naked Science*). The internal feature of modern sciences responsible for their successes was a matter debated throughout the twentieth century. It remains a compellingly controversial issue in the field of science education (Aikenhead, *Multicultural Sciences*). To be sure, there are contexts in which it will be important to distinguish between the practices of Western researchers and those of researchers in other cultures. But we will not take the exceptionalist route in doing so.

5. PREVIEW

Part I looks at three innovative and influential accounts in the field of science and technology studies, each of which has taken up the issue of modernity. These are the accounts of the French anthropologist of science Bruno Latour, the German sociologist and environmental theorist Ulrich Beck, and the European team of sociologists headed by Michael Gibbons, Helga Nowotny, and Peter Scott. (I shall refer to this last team as a single author with the initials GNS.) All three are unusual in that they focus on how concepts and practices of "the social" and "the political" must themselves be transformed in order to transform the sciences into more competent knowledge-producers as well as into resources for democratic social relations. Taking on this double concern is a valuable and rare kind of project to find in the field of mainstream science and technology studies. Moreover, it is not easy to find social theorists and political philosophers interested and courageous enough to have entered the world of science and technology studies in order to examine how to transform modern sciences and technologies for the kinds of politically progressive ends they recommend.

I also selected these three because they represent three distinctive sub-fields in science and technology studies, each with different resources to bring to the project of rethinking Western modernity's sciences and their social relations. Latour co-produced one of the early influential ethnographies of the production of scientific knowledge in *Laboratory Life* (Latour and Woolgar). His subsequent work has repeatedly raised important questions about the nature of scientific inquiry and the adequacy of prevailing philosophies of science. Beck approaches contemporary issues about science and technology from his work in the German Green Movement and from critical engagement with the tradition of sociological theory in which issues of modernity and modernization have always been centered. Indeed, nineteenth-century sociology was constituted by attempts to understand modernization's processes and effects in Europe—an origin which Beck, as well as others, suggests makes it an unlikely resource for thinking past modernization's conceptual framework. The GNS team's studies originated in a Swedish science policy assignment, and their analyses remain couched in terms useful to policy. Central concerns of Beck and GNS lie outside the ways the field of science and technology studies has come to define itself (cf. Biagioli, *Science Studies Reader*).[11] Yet GNS, and Beck to a lesser but still significant extent, are familiar with the central tenets of the conventional science and technology studies movement and situate their own work in the contexts of post-Kuhnian social studies of scientific and technological practices and philosophies.

While all three are critical of (Western) modernity and its philosophies and effects, they all also turn away from postmodernism as a solution to the crises of modernity. All three find it a valuable symptomology of modernity, but lacking a vision forward. In contrast to postmodernism, all three are optimistic about the possibilities of changing how science and politics are currently organized. Their activist, engaged stance toward how science is done is also valuable and rare in this field. They bring important resources to the projects of feminist and postcolonial science studies and to others grappling with issues about modernity beyond those raised by postmodernism. Yet each of these three accounts has severe though illuminating limitations, as we shall see in Part II. Neither the insights of feminism nor those of postcolonialism are engaged in these narratives. This failure undermines the potential success of their transformative projects.

Thus chapters 4, 5, and 6 look at the strengths and limitations of three fields of science and technology studies frequently ignored or misevaluated by mainstream progressive modernity and science studies and even by each other—feminist and postcolonial work on sciences and technologies, and their distinctive and illuminating ways of intersecting in the feminist work that is set in the context of the postcolonial analyses. In the course of their accounts, central dogmas of mainstream modernity theory are challenged. Each offers valuable strategies for transforming sciences and politics, North and South, to be more epistemologically competent and of use for pro-democratic projects.

In Part III, chapters 7 and 8 specifically look at ways that postcolonial and feminist science studies directly address modernity issues. Each of these modernity studies has significant stakes in demobilizing how modernity has been conceptualized as independent of and in opposition to tradition and how that oppositional contrast has been deployed in science, science studies, and public policy. Each offers visions of transformed sciences and politics which move past the modern impasses.

The litany of problems with modernity identified in the preceding chapters can be discouraging and immobilizing. The concluding chapter addresses the question of what can be done to transform the modernities that exist around the world today into ones which can and do deliver social progress to all of the world's citizens rather than only to an elite few. This is obviously a gargantuan task, and not one to which thinkers from the West alone could possibly have the best answers. Yet we will have seen diverse and valuable resources for such a project provided by Northern science and technology studies, on the one hand, and the postcolonial and feminist accounts, on the other hand. Each set of resources can enrich the other projects. In conclusion, I outline a modest proposal for obstructing the way that the modernity vs. tradition binary shapes research projects. Now we can turn to the innovative accounts of Northern science and technology studies scholars.

I.

PROBLEMS WITH MODERNITY'S

SCIENCE AND POLITICS

Perspectives from Northern

Science Studies

1.

MODERNITY'S MISLEADING DREAM

Latour

— — — —

FEW NORTHERN SCIENCE STUDIES have focused critical attention on the modernity of modern sciences. Innovative as these studies have been in undermining central foundations of still prevailing exceptionalist and triumphalist philosophies of science, modernity remains as a kind of horizon in this field. It restricts analysis to Western sciences and technologies and leaves the most powerful arguments of feminist and postcolonial analyses as seemingly unintelligible or irrelevant to science studies projects. In this respect, Western scientific and technological research remain understood in large part as positivism understood them, namely, as the complete terrain of what should count as scientific rationality and technological expertise. Indeed, by referring to them as "Northern" here, I intend to delimit the field on which we focus to the studies, whoever their authors may be, constrained by this kind of horizon.[1]

However, within the field of those who keep their gazes within this horizon, there are beginning to appear some critical foci on how problematic the modernity of the field so circumscribed is—on the modernity of Western sciences and technologies. This chapter and the next two will consider arguments by three Northern critics of modernity who have provided distinctive and extended analyses focused on the natural sciences and their philosophies. These are the French ethnographer and philosopher of science Bruno Latour (*We Have Never Been Modern*; *Politics of Nature*), the German sociologist of "risk society" Ulrich Beck (*Risk Society*; *Reinvention of Politics*; *World Risk Society*), and the team of European sociologists of science headed by Helga Nowotny, Peter Scott, and Michael Gibbons (Nowotny, Scott, and Gibbons, *Re-Thinking Science*; Gibbons et al., *New Production of Knowledge*). Each provides a distinctive focus, yet their arguments overlap in important respects. Let us begin by considering each of these features in turn.

First, each represents a particular influential focus in the field of contemporary postpositivist science studies. Latour's early study with Steve Woolgar, in *Laboratory Life*, of the social construction of results of research ("truth") in a biochemical laboratory was perhaps the earliest of the ethnographic studies of Northern sciences which so powerfully shaped the field of mainstream science studies. His subsequent interventions have again and again redirected the conceptual practices and debates within the field. Another important concern relevant to this project is with the ways scientific projects advance only by extending their technoscientific and bureaucratic networks greater and greater distances from their "centres of calculation" (Latour, *Science in Action*). Latour insists that attempts to explain scientific successes in terms of only their social causes are no more accurate than the older attempts to explain such successes in terms only of nature's order.

Beck's earlier studies were focused on the sociology of work. However, he is familiar with the findings of at least some of the post-Kuhnian work in the history and sociology of science, and specifically finds affinities in Latour's work with his own projects. Beck's work is in the lineage of classical sociological theory in ways Latour's (and the work of the Gibbons, Scott, and Nowotny group) is not. Beck also approaches issues of science and modernity from his activist experiences with the German Green Movement. Along with Anthony Giddens and Scott Lash, his work on "risk society" has been influential in Europe and North America (Beck, Giddens, and Lash, *Reflexive Modernization*). Most mainstream science studies scholars would probably not consider Beck part of their field since he does not do the laboratory or field site studies (or the equivalents in the history of science) which have come to dominate the field. Yet, as we will see, his work is highly pertinent to understanding what happens in laboratories and field sites.

Gibbons, Scott, and Nowotny, whose original study was commissioned by the Swedish government to aid in its science policy planning, are concerned especially with sociological and philosophical implications of the new ways in which European and North American sciences and technologies are being organized and practiced since the end of the Cold War in 1989. This work is in the lineage of science policy studies. So these three Northern science studies projects represent different approaches to rethinking modernity's sciences and politics.

Their arguments also overlap. First, while all are severe critics of modernity, its philosophies, and its effects, they all find postmodernism an unattractive alternative. For all three of them, postmodernism remains a valuable symptomology of problems with conventional thinking about modernity, but is stuck there. They each think that it does not have the intellectual or political resources to move beyond that critique in order to generate a positive program to transform modernity and its sciences. Thus their arguments demonstrate that postmodernism is not necessarily the inevitable landing site of critics of Western modernity.

Second, all three argue that (Western) science has become a kind of governance which illegitimately bypasses democratic processes. Thus "the scientific" and "the political"—science and politics—are inexorably intertwined. Science appropriates to itself as merely technical matters decisions that are actually social and political ones. However, a democratic ethic requires that everyone affected should participate in such decisions about how we will live and die—about which groups will flourish and which will lead nasty and short lives. On the other hand, social and political institutions constantly appeal to nature and to science to justify their own anti-democratic projects. The sciences we have and their philosophies intrude on and block possibilities for democratic governance. At the same time, the governance we have obscures its own intrusion into and permeability by authoritarian representations of the natural and the scientifically expert.

However, third, in contrast to many other critics of modernity and of modern sciences, all three are optimistic about the possibilities for transforming the sciences to be politically accountable for their practices and consequences. All three think the sciences can indeed contribute to social progress, but not without transformation of both the

sciences and the political worlds with which they are in mutually constitutive relations. All three call for more science, though they want
different kinds of sciences than those favored in the contemporary
West. They each strategize about how to democratize science in the
service of a democratized social order, and how to do so by strengthen-
ing and expanding the reach of the scientific impulse. So all three
overtly insist that science and politics—the scientific and the political
(or, in the case of Gibbons et al., the social)—must be simultaneously
redesigned. Thus all three raise issues about necessary transformations in both philosophies of science and political philosophies (or
social theories). This set of commitments and projects makes it difficult to categorize any of them as having fully modern or fully anti-
modern commitments. Indeed, as we will see, all three figure out ways
around that binary. And it sharply delineates their projects from the
vast majority of those in science studies which do not so overtly take on
the ambitious and controversial task of redesigning the realm of the
social and the political.

There are two more features unfortunately shared by all three. They
are all significantly gender-blind, and blind also to the analyses and
projects of postcolonial science studies, or at least to the most important features of the postcolonial accounts. Latour's accounts are at least
perched on the near side of the border between Western and postcolonial histories, ethnographies, and philosophies of science, without
fully appreciating the content or power of the postcolonial criticisms of
the West, its imperial sciences, and its modernities, let alone what
other cultures can offer Western sciences. With respect to gender, he
very occasionally does mention one or two feminist science theorists;
Donna Haraway gets perhaps three or four mentions in the two books
to be discussed here. Valuable as her work is, such a tiny citation record
is not sufficient to count as engagement with feminist science studies.
His work is uninformed by Haraway's arguments or those of any other
feminist science theorist. He specifically discounts the value of what he
refers to as "identity politics," including many of the new social movements which have produced feminist and postcolonial science studies.
Beck does gesture toward the welcome influence of women's movements globally. And he does venture into a topic put on the agenda of
social theory by feminism—"Love" (Beck and Beck-Gersheim, *Normal
Chaos of Love*). But, like Latour, he does not actually discuss or take
account of issues produced by feminist and postcolonial science stud-

ies. Indeed, he does not even mention, and I would guess is unaware of, the latter. He does not try to connect to his analysis of science and politics the fascinating account provided with Elisabeth Beck-Gersheim of the relation between quandaries of intimate relations and recent changes in the organization of work and family life. To be sure, this would be an ambitious project, but we will see it at least partly engaged by feminist modernization accounts. Nowotny, Scott, and Gibbons do mention feminism once or twice, but do not even achieve a "gesture" toward the issues feminist science studies have raised. They appear totally unaware of postcolonial science studies.

Thus the accounts of these three influential theorists forge ahead as if the backs of feminists and others "excluded" have been glimpsed retreating over the horizon of modernity into their natural worlds of "tradition," but leaving no traces behind in conceptions of modernity or its ideal social relations. To these scholars, like so many others in the field of science studies, feminist and postcolonial experiences and analyses produce no relevant insights or strategies which could or should change the way these otherwise innovative authors conceptualize and carry out their projects to transform scientific inquiry and the domain of the social and the political. To them such experiences and analyses appear irrelevant, even incomprehensible or, in the case of Latour, as largely obstacles to scientific and social progress.

In this respect their work continues an unfortunate tendency. Mostly invisible to them, but not to feminists and postcolonial theorists and researchers, are the long histories and present projects of male supremacy and imperialism/colonialism, hulking like two proverbial 800-pound gorillas in the parlors, parliaments, board rooms, and laboratories of modernity and its sciences. ("Mostly" invisible, since Latour does get brief glimpses of at least the colonialism gorilla!) In ignoring androcentric and Eurocentric aspects of both their objects of study and their own accounts, all three deeply undermine the epistemic and political chances of success of their own projects. The legitimacy of speaking in the voice of the ruling, white, Northern, bourgeois, "rational man" has radically declined everywhere around the globe. Those of us concerned with gender and postcolonial social justice need the projects of these theorists as well as our own to succeed, so these lacunae require attention. To the extent that they distance their accounts from feminist and postcolonial analyses, they inadvertently disable their own projects and end up functioning as support for the manipulation of male-supremacist and Eurocentric anxieties for the ends of social *in*justice.

Yet, it must be emphasized, these three theorists do raise important questions about modernity and its sciences and philosophies which have not yet been centered in feminist and postcolonial work, and they provide valuable insights into illuminating ways to conceptualize issues which feminist and postcolonial work has thought about only in other terms. Feminist and postcolonial science and technology studies and Northern science studies scholars can learn from each other.

Let us turn to Latour's criticisms of modernity to see one account of the resources that a critical view of both modern science and modernity's politics can provide for epistemological and democratic transformations.

1. HAVE WE EVER BEEN MODERN? ONTOLOGICAL PROBLEMS

In a series of books and articles Latour has conducted a vigorous campaign against conventional and even some postpositivist epistemologies and philosophies of science. He proposes a radical alternative to them which, in addition to its other virtues, in important respects better fits the actual practices of the natural sciences. Latour's arguments are innovative, rich, dense and impossible to do justice to in a brief report. I shall capture some central themes of his joint revision of dominant notions of science and of politics by following his argument that we can create a far sounder and more pro-democratic conceptual framework than that provided by the fact/value distinction and its correlates. For Latour, the fact/value distinction also underlies such binaries as scientific analysis vs. social analysis, science vs. technology or applications of science, nature vs. the social or the political, objective vs. subjective, rational vs. irrational, and modern vs. premodern.

In his influential book *We Have Never Been Modern*, Latour argued that modernity and its sciences have an ontology problem. They conceptualize our knowledge of nature as separate from matters of our interests, of justice, and of power, though it is in fact inseparable.[2] "On page six [of my daily newspaper], I learn that the Paris AIDS virus contaminated the culture medium in Professor Gallo's laboratory; that Mr. Chirac and Mr. Reagan had, however, solemnly sworn not to go back over the history of that discovery; that the chemical industry is not

moving fast enough to market medications which militant patient organizations are vocally demanding; that the epidemic is spreading in
sub-Saharan Africa. . . . [H]eads of state, chemists, biologists, desperate
patients and industrialists find themselves caught up in a single uncer-
tain story mixing biology and society" (1–2). We live in an incommen-
surable mix of nature, politics, and discourse. "Yet no one seems to
find this [story] troubling. Headings like Economy, Politics, Science,
Books, Culture, Religion and Local Events remain in place as if there
were nothing odd going on. The smallest AIDS virus takes you from sex
to the unconscious, then to Africa, tissue cultures, DNA and San Francisco, but the analysts, thinkers, journalists and decision-makers will
slice the delicate network traced by the virus for you into tidy compart-
ments where you will find only science, only economy, only social
phenomena, only local news, only sentiment, only sex. . . . By all
means, they seem to say, let us not mix up knowledge, interest, justice
and power. Let us not mix up heaven and earth, the global stage and the
local scene, the human and the nonhuman. 'But these imbroglios do
the mixing,' you'll say, 'they weave our world together!' 'Act as if they
didn't exist,' the analysts reply" (2–3).

Thus the world we experience consists of networks linking aspects
of nature, cultural legacies, states and nations, agencies, institutes,
corporations, official and unofficial policies, de facto practices, mechanisms and other artifacts, and even deities. The basic constituents of
the world are such hybrid networks. Yet modernity requires a representation of reality consisting only of images of purified objects. It
delinks nature from culture; appropriate policies and practices from
deities; agencies, institutes, and corporations from mechanisms and
other material artifacts. Modernity, its epistemologies, its philosophies
of science, and its sciences represent a world of broken networks and
dismembered hybrids—not the one in which we live or about which we
want explanations. Its sciences are intentionally isolated from the reality that needs explanation. Its conceptual framework leaves nature's
"constituents" with no "voice" in the politics with which they are in-
fused and which they are called on to justify. We do not live in the
modern world that the epistemologies and philosophies of science of
modernity imagine; indeed, we and our sciences have never been modern, Latour proclaims. We and our sciences are as historically specific,
as much in the thrall of an only-imagined reality, as any culture and its
knowledge system could be, he implies.

Yet we need sciences for that pre-modern world in which we live. Where, Latour asks, are sciences of these networks and hybrids which modernity, its epistemologies and philosophies of science refuse to conceptualize as such? Not in the laboratories. Rather, he argues, the sciences we need have been conceptualized and developed in the field of science studies (cf. Biagioli, *Science Studies Reader*). Its histories, sociologies, ethnographies, and textual studies of moments in the history of Western sciences focus on the relations between knowledge-seeking and the social, cultural, economic, political, and even psychic projects of an era. Indeed, they show how the latter get inside the cognitive core and content of even the very best of Western scientific achievements.

Latour does not wish to abandon the West's Enlightenment project, in contrast to some other critics of Western philosophies of science. Instead, he proposes that we redefine the Enlightenment to exclude its narrow and distorting vision and practices of modernity. Thus he is opposed to both the illusion and the ideal of modernity. So, too, are postmodernisms. However, like the other figures we will examine, he thinks that postmodern analyses simply give up and enjoy the confusion of the present moment without trying to improve the sciences we have or their faulty ontologies.[3] Latour proposes that we think instead in terms of the nonmodern or amodern (*We Have Never Been Modern* 47). Both the need for more competent sciences and the political work of constructing democratic social relations require reuniting aspects of the world which modernity keeps sundered. "Half of our politics is constructed in science and technology. The other half of Nature is constructed in societies. Let us patch the two back together, and the political task can begin again" (144). He argues that we need a "parliament of things" to participate in political decisions about nature's simultaneously natural and social properties.

Thus Latour binds to each other failures of the modern political project and failures of its knowledge project. Advancing democratic social relations as well as restoring the environment, another important political project, both require a scientific study of kinds of objects around us that modernity has banished from view. It is the recently emerging field of science studies that has developed the resources to engage in such work, he argues. Such studies bring systematic scientific assumptions and methods to the description and explanation of the hybrids and networks which constitute reality.

2. AFTER FACTS VS. VALUES

In a later book, *Politics of Nature*, Latour returns to the project of re-distributing the still-useful aspects of the fact/value distinction in a different way while explicitly refashioning the political, a project to-ward which he only gestured in the earlier book. What does it mean, he asks, to understand nature, on the one hand, and the social and politi-cal on the other hand, as always already inside each other rather than as discrete territories?

Against the social construction of nature

By 2004, Latour is worried not only about the conventional philosophy of science which claimed that only nature and scientific method were responsible for the very best scientific claims; now he is also sharply critical of certain postpositivist science studies tendencies which seem to give *nature* too little credit for the legitimacy of scientific claims. Latour distances his project here from the "social construction of na-ture" way of thinking about the issue which is characteristic especially of the influential sub-field of science and technology studies known as the sociology of scientific knowledge. This conceptual framework leaves nature outside the domain of the social and political and focuses only on how humans interpret or represent this "outside" in social and political terms (*Politics of Nature* 32–42). Thus it "abandons to Science and scientists" the study of what actually happens in nature (33). Na-ture must be given a more important role in the production of sci-entific knowledge than the social constructionists seem willing or able to do.

The social constructivists support mere multiculturalism, which, he says, settles for a relativist conceptual framework of incommensurate worlds wherein different cultures simply have different beliefs about nature and social relations. "Multiculturalism acquires its rights to multiplicity only because it is solidly propped up by *mononaturalism*" (33). That is, Latour is saying, multiculturalism is grounded in the commitment to a natural order that is a priori fixed and thus with-out any social or political components. Such an understanding makes judgments impossible about which beliefs are the better ones, and about which is the best possible world for us to live in. Such multi-culturalism makes us unable to "compose a common world." By this phrase Latour seems to mean a completely common world, such that we and others can always understand exactly what the other means—a point to which we return. His own account is all too often read in one of these problematic ways, he claims, though he is arguing something quite different.

Plato is the originary culprit

Latour locates the origins of the problematic fact/value distinction in Plato's myth of the cave. Plato insisted on a nature free of social and political elements and a politics with purportedly no connections or effects on nature—that is, on reality. The cave myth is still useful, Latour argues, because it permits the accumulation of power by the institution of Science. (Latour distinguishes Science as the institution, its practices, and culture as these appear in modern philosophies from the sciences that are actually practiced today.) That is, we could say though Latour does not put it this way, the continuing usefulness of Plato's myth and the fact/value distinction is to be found in the interests of scientists and their institutions.

The continuing value of the model results from the role played in it by just a few people, the philosopher-scientists who are capable of traveling between the cave dwellers and the outside world and thereby "converting the authority of one into that of the other. . . . the Myth of the Cave makes it possible to render all democracy impossible by neutralizing it; that is its only trump card" (*Politics of Nature* 13–14). It is scientists (and their philosophers) who have illegitimately hijacked political powers. "These few elect, as they themselves see it, are endowed with the most fabulous political capacity ever invented: *They can make the mute world speak, tell the truth without being challenged, put an end to the interminable arguments through an incontestable form of authority that would stem from things themselves*" (14; original italics).

Latour proposes that this explains why scientists have been resistant to science studies, which have from the beginning challenged the authority of scientists alone to provide legitimate and accurate accounts of the natural world:[4]

> What might political philosophy look like if we abandoned the allegory of the Cave? How can we conceive of a democracy that does not live under the constant threat of help that would come from Science? What would the public life of those who refuse to go into the Cave look like? What form would the sciences take if they were freed from the obligation

to be of political service to Science? What properties would nature have if it no longer had the capacity to suspend public discussion? Such are the questions that we can begin to raise once we have left the Cave en masse, at the end of a session of (political) epistemology that we notice retrospectively has never been anything but a *distraction* on the road that ought to have led us to political philosophy. Just as we have distinguished Science from the sciences, we are going to contrast power politics, inherited from the Cave, with politics, conceived as *the progressive composition of the common world.* (*Politics of Nature* 18)

Fact/value strengths

One point we must clarify about Latour's argument is that he identifies the strengths of the fact vs. value distinction and then figures out how to preserve them while jettisoning the fact vs. value distinction itself. The most important task of this distinction is to guard against ways that ideologies come to shape scientific practices.

> If we were to show, for example, that immunology is entirely polluted by war metaphors, that neurobiology consumes principles of business organization in enormous quantities, that genetics conceives of planning in a determinist fashion that no architect would use to speak of his plans, we would be denouncing a number of frauds used by smugglers to conceal debatable values under the umbrella of matters of fact. [He cites Keller on genetics.] Conversely, if we were to denounce the use a political party makes of population genetics, or the use novelists make of fractals and chaos, or the use philosophers make of the quantum uncertainty principle, or the use industrialists make of iron-clad economic laws, we would be denouncing the smugglers from the other side who hide under the name of Science and sneak in certain assertions that they dare not express openly, for fear of shocking their public, but that obviously belong to the world of preferences—that is, values. (*Politics of Nature* 99–100)

The fact/value distinction guards against permitting the judges of the adequacy of a scientific trial to be those people with interests in how the trial comes out. We need to redistribute the properties of facts and values into different bundles. The present notion of a "fact" obscures the process of fabricating facts and the role of theory or paradigms in making data coherent (96). Meanwhile, Latour says, pity the poor values which always arrive on the scene only after the facts have already taken up their projects. "Values always come too late, and they always find themselves placed, as it were, ahead of the accomplished fact, the *fait accompli*. If, in order to bring about what ought to be, values require rejecting what is, the retort will be that the stubbornness of the established matters of fact no longer allows anything to be modified: 'The facts are there, whether you like it or not.' . . . Once the cloning of sheep and mice has become a fact of nature, one can, for example, raise the 'grave ethical question' whether or not mammals, including humans, should be cloned. By formulating the historical record of these traces in such a way, we see clearly that values fluctuate in relation to the progress of facts" (97).

Moreover, "even if they reject this position of weakness that obliges them always to wait behind the fluctuating border of facts, values still cannot regroup in a domain that would be properly theirs, in order to define the hierarchy among beings or the order of importance that they should be granted. They would then be obliged to judge *without facts*, without the rich material owing to which facts are defined, stabilized and judged. The modesty of those who speak 'only about facts' leads astray those who must make judgments about values. Seeing the gesture of humility with which scientists define 'the simple reality of the facts, without claiming in any way to pass judgment on what is morally desirable,' the moralists believe that they have been left the best part, the noblest, most difficult part!" (97). In short, "the common world and the common good find themselves surreptitiously confused, even while remaining officially distinct" (98).

Designing a new "inside" and "outside" of science

Latour thus calls for a different "separation of powers" (*Politics of Nature*, chap. 3) from the one between the authority of scientists to pronounce on facts and of "moralists" and politicians to pronounce on values. Here Latour reconceptualizes the "inside" and "outside" of science and of society (e.g., 121–25). In the modern account, only pure facts, primary qualities are permitted on the inside of science. All other social, cultural, and political values and interests, the secondary qualities, are to be kept outside science, contained out in society. But in Latour's account, the inside contains whatever humans, animal species, research programs, concepts, or other phenomena for which the society ("the collective") does take responsibility through exercising the due processes of "taking into account." The outside contains two kinds of exteriorities; one is the rejected entities that the society refuses to take responsibility for through the "power to put in order": "Of these excluded entities we cannot say anything except that they are exteriorized or externalized: an explicit collective decision has been made *not* to take them into account; they are to be viewed as insignificant. This is the case . . . [for] the eight thousand people who die each year from automobile accidents in France: no way was found to keep them as full-fledged—and thus living!—members of the collective. In the hierarchy that was set up, the speed of automobiles and the flood of alcohol was preferred to [eliminating] highway deaths. . . . [T]hese entities can be humans, but also animal species, research programs" (124). They can be any of the other phenomena "that at one moment or another are consigned to the *dumping ground* of a given collective" (124). Yet such excluded entities always "*are going to put the collective in danger,*" for they can return to "haunt the power to take into account" at any time (124, 125). Thus Latour insists that no entity can ever be confidently permanently banished to the outside of science/society, and that threat of return is an important force for maintaining "due process" in the constitution of democratic social relations in science and in society. Here Latour tries to build in a "feedback loop," a check and balance to inspire far more careful and accountable processes of deciding whose "voices" to cultivate in designing and managing nature/society through science/politics. "The collective" (formerly referred to as "society") contains whatever humans agree to conceptualize and knowingly interact with. Whatever they refuse to "take responsibility for" in such ways is banished from "the collective."

The second kind of exteriority consists of those entities and phenomena which have not yet come to the attention of society (191ff.). For example, until recent decades, tectonic plates, AIDS viruses, and moon rovers were also part of the exterior of science and society, as we discovered once they entered the interior by appearing in scientific accounts and in social thought.

Latour then elaborates in subsequent chapters the scientific and political conditions and effects of such a reconceptualization. Through metaphors of a parliament with "two houses," he tries to articulate just how democratic processes would direct a society's always simultaneously scientific and political projects. For the purposes of our project, and for reasons which will shortly become clear, we do not need to pursue Latour's argument here further.

3. LIMITATIONS

Latour's project gives us much to ponder. It raises important questions for anyone interested in thinking about transforming Western sciences and their political worlds. It has other important resources to which I return in the conclusion. He is still at work on it so it would be premature to make any conclusive statements about it. Yet its limitations so far are significant and instructive. Because these limitations are grounded in some of his fundamental assumptions and commitments—they are not accidental—one can wonder if Latour would ever be willing to engage with them. They can be summarized under two headings. First, while his account overlaps with feminist and postcolonial concerns in important respects, he does not think it important or valuable to engage with these movements' analyses even when they focus on precisely some of the issues of greatest concern to him. He misunderstands these movements' projects, at least in part because of the way he conceptualizes desirable forms of politics. Thus he cannot hear how his justification for ignoring them can create, at best, only a "palace coup," a changing of the guard, not the kind of far-reaching political revolution he sometimes seems to desire and think he is guiding. Second, his political and social theory, including his conception of the modern, is not only thin (as he notes); it is also headed in a direction counter to the interests and desires not only of feminist and postcolonial groups, but also of other pro-democratic social movements. I take up these two kinds of limitations in turn.

First, Latour fails to engage with feminist, anti-racist, or postcolonial criticisms of either the natural sciences and their philosophies or of political philosophy.[5] This is not just an oversight. Rather it is a principled non-engagement on his part, and it has negative consequences for his account. It originates in assumptions that lead to misrepresentations of these movements and their claims. "The lower house [in the new kind of democratic government he proposes] asks the question 'Who are we?' This 'we' is variable in its geometry; it changes with every iteration. Unless we are dealing with repetitive collectives that already know, have always already known, of what they are composed— but these collectives, whether on the right or the left, whether based on racial identity, the nature of things, humanism, or the arbitrariness of the sign, do not belong to the realm of political ecology. They all stem from the Old Regime, since, for them, two distinct domains of reality

order all facts and all values in advance. Their metaphysics is not experimental but identity-based. We are only interested here in the collectives whose composition is going to be modified with each iteration
—even if they have to reinvent themselves in order to remain the
same" (*Politics of Nature* 173).

Identity politics

To the contrary, the metaphysics of these groups is always necessarily
experimental, as Latour puts the point. The pro-democratic social
movements now choose to claim identities, to name themselves, precisely against the way the Old Regime defined them as objectively
a priori determined groups in themselves and then treated them in-
equitably—as not fully human. The Old Regime invoked as scientific
fact their identities as inferior groups of people. They were always
only immature, damaged, or deviant forms of the ideally human.[6] It
brought them into existence only as social groups "in themselves"—as
the inevitable poor, or as "colored," "savages," or "primitives," women,
"queer," or "underdeveloped." As such they became suitable objects
for revealing their a priori nature to natural and social science research
and to the ministrations of public policy which the natural and social
sciences serve.

But surely Latour understands this much. What he does not seem to
grasp is that these "identity movements" bring such groups into existence in a different way, namely, "for themselves," as self-conscious
collective agents of history and knowledge. Generating thought from
the conditions of their lives enables both them and us to understand
the dominant institutions, as well as their conceptual and material
practices of power, in critical ways which are hard to detect from the
perspective of the dominant conceptual frameworks and material practices. Indeed, Latour's principled rejection of such standpoints damages his own project, as we shall see.

Moreover, Latour misses the sociological insights about the function
of social identities. Sociologists look at identities as ways in which civil
society gives meaning to social life (Castells, *Power of Identity*). The
influential identities of industrializing societies were those of class
("working men" or "educated classes"), religion (Catholic, Protestant,
Jewish), nationalism, and political parties. Were these "pre-given," as
Latour charges? Yes and no. Individuals certainly tended to grow up
knowing they were, for example, Irish, Catholic, workers who voted

Labor, or American, Protestant, educated people who joined Lincoln in voting Republican. Yet each of these identities was continually expanding or contracting as, for example, immigrant groups became American; higher education was vastly expanded after World War II; new fundamentalist, pentecostal, and charismatic Protestant denominations drew worshipers from older denominations; and many well-educated people began to support the domestic policies of Democrats such as Roosevelt, and then of Kennedy and Johnson. These classical social identities were always in flux, continually "recomposing" themselves in the face of changing social relations and perceptions of them.

To be sure, some people do take their proudly reclaimed identities as African American, women, or queer to be pregiven by biology or even by history. Yet many Americans with African American ancestry have chosen to identify themselves as white, and many who look white are choosing to identify as African American or as Black. Thus Patricia Hill Collins and other African Americans insist that their Black feminism and/or nationalism is a political commitment, not a description of their biology (Collins, *Black Feminist Thought*, chap. 1). Similarly, people in other groups have insisted that their political projects are chosen, not pre-given by nature or society. Latour seems to conceptualize these groups through the old, long-discredited concept of cultures as isolated, timeless, and static communities speaking languages only they understand. I doubt that Latour actually could think such cultures exist anywhere now or in the past. Indeed, it is precisely the contrast between Western Science and the thought of such imagined non-Western cultures which Latour adroitly undermines, intentionally or not, in *Science in Action* and, by implication, in the very title of *We Have Never Been Modern*.

Grounds for identities or for truths?

In a related way, Latour confuses the grounds for claiming such identities with the grounds for the truth or empirical adequacy of the claims that such groups make. Such identities are claimed typically because of the perception of shared cultural and political histories and thus shared political projects today. But the truth or empirical adequacy of the claims such movements make depends, such movements often point out, upon broadly familiar epistemological standards. Can the claims gather better empirical support than their competitors? Do they explain phenomena of interest to such groups which were ignored by

their competitors? These groups insist that there are natural and social realities about which they need to know in order to improve the
health, environment, or economic, political, and social relations, or
other conditions of their lives. They ask different questions, and come
up with different answers to familiar questions. To be sure, the new
pro-democratic social movements have challenged conventional standards for knowledge in many ways. They have introduced novel kinds
of knowers (those "inferior" peoples), kinds of evidence (their perceptions, and the view from their everyday lives), and ways of understanding objectivity and good method (see chapters 4, 5, and 6). Such groups
have indeed made such revisions in science and politics, but they
have argued for them to people and institutions without such identities, such as research-funding organizations, publishers, tenure com-
mittees, legislatures, and national and international legal, economic,
health, education, and other such institutions. They have behaved just
as Latour does when faced with excessive constructivist understandings of his projects by science studies.

To be sure, some advocates for the rights of women, Native Americans, Latin Americans, or Muslims have also suffered from this confusion, claiming a final authority for the truth or empirical adequacy of
their perspectives on their lives and social relations more generally that
cannot legitimately be awarded them. But enough do not, and in fact
enough criticize the idea that gaining experience and gaining knowledge are identical processes to make Latour's wholesale dismissal of
"identity politics" untenable. These groups, too, recognize that others
—economists, historians, or therapists, for example—often can have a
broader or more objective view than they do about just what they
experience, and how it reasonably might best be described. Moreover,
we often change our own views of what we have experienced, as a book
with the title of *I Never Called It Rape* (Warshaw) reveals. Such groups
often have to invent new concepts to draw attention to their concerns;
this was the case for feminist inventions of such concepts as sexual
harassment, women's double day of work, and sexual politics. These
phenomena were not comprehensible through the available ways of
thinking about gender relations. After all, social science would be impossible if everyone had the final word about the adequacy of his or her
own experiential reports. Yet the social groups Latour externalizes (to
apply his terminology to his practices) have produced and stimulated
in others some of the most startling and influential social science

findings of recent decades, as even his gestures toward them seem to indicate. My point here is that the grounds for such groups forming as groups are different from the grounds for accepting or rejecting claims that they make, and this is well understood by many influential thinkers in these groups. Latour misrepresents their practices here, to the detriment of his own project.

Locating common ground

Latour seems overly anxious about the difficulty of locating or constructing the "common ground" needed for discussion of how to transform science and politics in the face of their failures. He seems to want a unity which he can define and to which no one will object on political grounds. "The only way to compose a common world, and thus to escape later on from a multiplicity of interests and a plurality of beliefs, consists precisely in *not* dividing up at the outset and without due process what is common and what is private, what is objective and what is subjective. Whereas the moral question of the common good was separated from the physical and epistemological question of the common world, we maintain, on the contrary, that these questions must be brought together so that the question of the *good* common world, of the *best* of possible worlds, of the *cosmos*, can be raised again from scratch" (*Politics of Nature* 93). But why on earth should one want, let alone expect to find, a way to escape from a "multiplicity of interests and a plurality of beliefs"? This seems to be a displacement of reasonable anxiety about the appropriate authority to claim for his own account, a point to which I turn shortly. Yet Latour focuses this anxiety on the purported impossibility of communication between, on the one hand, such groups as feminists, postcolonials, or African Americans, not to mention African Muslims in France, and, on the other hand, individuals who do not claim such identities. Shouldn't sciences, politics, and their philosophies be addressing how to create, instead of fantasies of falsely-grounded unities, societies which can recognize the necessity of differing interests and beliefs about nature and social relations and yet create pluricentric and democratic modes of getting on together?

Part of the problem is that for Latour, gender, race, and ethnicity seem to refer only to individuals, or "collectives" of them, and only to individuals of the marked category (in spite of his praise for feminists' criticism of this latter tendency). He admits that he, too, happens to have a gender and race, not to mention an avowedly deep commitment to the "glorious Republican legacy" of France (*Politics of Nature* 165). But these are simply historical accidents awaiting a global democratic society's working them over into a common world. "Once we have exited from the great political diorama of 'nature in general,' we are left with only the banality of multiple associations of humans and non-humans waiting for their unity to be provided by work carried out by the collective, which has to be specified through the use of the resources, concepts, and institutions of all peoples who may be called upon to live in common on an earth that might become, through a long work of collection, the same earth for all" (46). Yet Latour's preoccupation with attaining unity and a common world blinds him to the cognitive and political importance of differences in the worlds in which we live. It is precisely the identity movements which have been able to develop a "principled relativism" (in Fredric Jameson's phrase in "'History and Class Consciousness'")—a way to leverage the uncommonness of our shared worlds in order to advance both the growth of knowledge and of democratic politics. It does not follow from acknowledging others' differences from ourselves that we will not be able to communicate with each other and, indeed, agree on a great deal, contrary to Latour's fears. It does follow, however, that we will have to listen carefully to each other in order to understand ourselves and our worlds, as well as to understand them and their worlds, even if only partial understandings are possible of either. For one thing, Latour will have to take more seriously the point feminists were making in criticizing the way men and masculinity tend to generate distinctive assumptions and beliefs. They are not capable of producing universal truths contrary to dominant ways of thinking.

Postcolonial issues

At this point we must introduce another of Latour's important analyses, namely, the discussion of how the "facts and machines" of technosciences succeed only when they can travel on ever-extended social/cultural/material networks. Though this account is prefigured in even earlier work, it is fully elaborated in *Science in Action* (especially chapter 6), and a full "case study" appears in his discussion of the immense administrative/managerial labors Pasteur undertook to actually "do the science" for which he is so famous (*Pasteurization of France*). It would take us too far afield to take the space here to do justice to this splendid work. There are just a couple of points I want to make about this analysis.

In *Science in Action*, Latour directly challenges the all too familiar idea that other cultures have only local beliefs about nature and social relations, while modern societies gain universal knowledge about them through their sciences. Latour argues that what distinguishes modern from premodern societies (he does not use these designations) is rather that the modern societies alone are capable of building gigantic networks of cultural, social, and material kinds of persons, processes, and—most importantly—"inscriptions" on which the advance of modern sciences can travel rather like trains on rails. Such networks permit the emergence of "centres of calculation" from which sciences learn "how to act at a distance on unfamiliar events, places and people" (223). This is achieved "by *somehow* bringing home these events, places and people. How can this be achieved, since they are distant? By inventing means that (a) render them *mobile* so that they can be brought back; (b) keep them *stable* so that they can be moved back and forth without additional distortion, corruption or decay, and (c) are *combinable* so that whatever stuff they are made of, they can be cumulated, aggregated, or shuffled like a pack of cards. If those conditions are met, then a small provincial town, or an obscure laboratory, or a puny little company in a garage, that were at first as weak as any other place will become centres dominating at a distance many other places" (223). In subsequent discussion in the same book, Latour continually intersperses examples of how this happened through the European "voyages of discovery" with examples drawn from other history of science sites, as in the passage above. He explicitly credits two of the few then available postcolonial accounts of the relation between the advance of European sciences and European imperialism and colonialism (225; Brockway, *Science and Colonial Expansion*; Pyenson, *Cultural Imperialism and Exact Sciences*). So it would not be accurate to say that Latour is completely unfamiliar with the postcolonial science studies literature. While not much of it was available in English or French in the early 1980s, Latour did locate a couple of significant examples.

Yet Latour seems completely unaware that the account he is developing—of how Western sciences gain power only as they "extend" the networks further and further from their (Western) centers of calculation, and of how important bureaucracies' "paper pushers" are (*Science in Action* 254ff.) to the advance of (Western) science—is also an account of how Western sciences and empires are inextricably inter-linked, how they co-constitute each other, and in disastrous ways for those "left on the outside" of the West's sciences and societies. More-over, he seems unaware that these other belief systems which West-erners encounter might have something valuable to offer Westerners. His focus is on how the networks on which Western technosciences travel are constructed and work, and not on the histories of the worlds into which such networks are extended before or after their encounters with the West. His metaphor of networks extending, so valuable in some ways, is problematic in others.

Now I come to the second kind of limitation in Latour's work. The cost to Latour's account of his principled refusal to engage with femi-nist and postcolonial studies is exacerbated by certain assumptions, practices, and directions in his social theory and political philosophy. The first thing to note is that Latour's foray into transforming notions of the political as he transforms notions of nature and science has the character more of a philosophical thought-experiment than of an informed scholarly analysis. That is, it makes no attempt to signal familiarity, let alone engage, with the issues political philosophers have been discussing about which conceptions and practices of democracy are most desirable and what the limitations are for different social groups (races, genders, classes) with respect to, for example, Liberal, Republican, socialist, participatory, and deliberative/communicative conceptions of democracy. He would probably agree that a thought-experiment was his goal. Perhaps it is unfair to ask for more, for where does one find philosophers or ethnographers of Western sciences who are also at home in the world of political philosophy? Or political philosophers at home in science studies? Yet this widespread separa-tion of such disciplinary expertise is surely part of the challenge of trying to simultaneously think about science and politics, as Latour so admirably wants to do.

Nor, as he makes clear, does he provide an activist's analysis that attends to the strategies likely to be most useful to move toward a particular ideal—his own, or others'. He would most likely accept this criticism, too, since he several times notes that his working out of the governance details of the new philosophy of science/politics is a "placeholder" for a more considered proposal.

However, in the second place, even if Latour's account is only a "placeholder," it is not a value-neutral one. Rather he makes clear a number of his commitments. He commits himself and his analysis to "the republican heritage of our ancestors" (*Politics of Nature* 165) in the very paragraph in which he insists that the language of "parliament," "upper house," and "lower house" are "here only to point out provisional sets of competencies, to allow the diplomats to speak" (165).[7] Valuable as central theses of republicanism may be, he does not consider the criticisms raised against it in recent defenses of liberal, participatory, and deliberative/communicative democracy (e.g., Benhabib, *Democracy and Difference*). These critiques are "externalized," in Latour's terms, from his ideal society; that is, these criticisms are not treated as significant actors in contemporary social thought and social relations.

Another set of externalized social actors are social movements, and, significantly for his project and ours, pro-democratic identity-based movements, as indicated above. These movements have already had important transformative effects on scientific programs, especially health and environmental programs, and now increasingly on economic policy of international financial institutions through, for example, the World Social Forums which now accompany meetings of the International Monetary Fund, World Bank, and other international economic agencies (cf. Mertes, *A Movement of Movements*). Other contemporary theorists engaged in projects related to Latour's, such as Ulrich Beck and Manuel Castells, see such movements as a promising political force for democratic transformation that exists in the world today, as I will discuss in later chapters. But Latour simply disvalues and then ignores them, assigning them to the "not real" in terms of recognizably and desirably influencing science or society.

A third problem here is that the modern on which Latour focuses seems to be only a set of ideas contained within the philosophy of science and political philosophy. When Latour proclaims, "We have never been modern," he seems to restrict his attention to the dominant philosophy of modern science. He makes scattered remarks about the "chaos" that characterizes modern political life, but does not get into focus a good deal of what defenders and critics alike refer to as modernity and as modernization theory. We get no sense of the focus of such theorists on such real social processes as industrialization and urbanization, let alone Third World development and its associated global restructuring processes.

Finally, lurking behind these other problems is the question of who is the "we" who have never been modern. The big news Latour's account brings is not that "we" have never been modern, but that bourgeois, Western men who get to construct philosophies of science and political philosophies have never achieved that status to which they aspired. The rest of us always suspected not only that women and the West's Others already were not modern; worse, we never would be, at least not in the ways the Enlightenment's ideal Rational Man could be. As it turns out, the assignment of the rest of us to pre-modernity has always been the prerequisite for Western bourgeois men's illusions of modernity, as later chapters will delineate.

Furthermore, the pre-modern, in which "we" have always existed contrary to our illusions of modernity, is only the object of his inquiry (and an extremely partial and abstract one at that), not his condition as speaker or ours as readers. Latour, the speaker, the subject of these works, is firmly located in the dream of modernity. He and his program appear to have no relevant historical specificity; no particular social changes lead to his reflections. He speaks as a unitary, centered subject providing an authoritative, socially neutral account (even if he does admit to a number of value preferences). There is little overt analysis of the social networks in which his own account is embedded, or awareness of the networks into which his writing will fall, that is, apart from its probable uptake in science studies. There is no awareness of the "conceptual practices of power" (Smith, *Conceptual Practices of Power*), in which his particular philosophic account is engaged, intentionally or not. Thus, paradoxically, he, the speaker, and we, the readers, are to remain in the modern mind-set as we read his text, in spite of our interest in hybrids and networks rather than pure facts or values. Latour has no sense of the difference which the absence or presence of women and other "minorities" does and could make in the production of scientific knowledge. Nor does he have a sense of the difference it could make in the production of the social studies of science and technology. He has no idea that such people, perceiving the world through the eyes of new social movements, would see quite differently the issue of "we have never been modern."

To be sure, Latour does try to conceptualize a relationship between speech and authority different from the one legitimated by modern philosophy of science. In the "consultation" stage of finding "common ground," he recommends a "slow search for reliable witnesses capable of forming a voice that is at once hesitant and competent" (112). "The meaning of the words 'discussion' and 'argument' is modified as soon as we evoke scientists in lab coats. . . . There are more and more common arenas in which discussion is nourished both by controversies among researchers and by squabbling in assemblies. Scientists argue among themselves about things that they cause to speak, and they add their own debates to those of the politicians. . . . It would thus be wrong to see people who do not discuss because they demonstrate— scientists—as opposed to people who discuss without ever being able to reach agreement on the basis of a definitive demonstration—politicians. Where are we going to find the means of buttressing, provisionally, this capacity of speech that is intermediary between 'I am speaking' and 'the facts are speaking,' between the art of persuasion and the art of demonstration . . . ?" (63–64). Here Latour begins to identify a central issue for feminist and postcolonial critics of modernity, one that we will pursue in later chapters. Latour also modestly insists that his is just one view, and one at least geographically local: "As I am well aware, I have expressed only one particular viewpoint, one that is not simply European but French, perhaps even social democratic, or worse still, logocentric. But where has anyone seen a diplomat who did not bear the stigmata of the camp he represents? Who does not put on the livery of the powerful interests that he has chosen to serve and thus to betray? . . . Am I therefore limited to my own point of view, imprisoned in the narrow cell of my own social representations? *That depends on what follows*" (221). And what follows must be negotiations. "Everything is negotiable, including the words 'negotiation' and 'diplomacy,' 'sciences' and 'democracy'—simply white flags waved at the front to suspend hostilities" (221–22). We readers can be glad to hear this. But one can't negotiate with the world of the nonreal to which Latour has already assigned "identity politics." What he does know, however, is that such ghosts can return to haunt the "real world." These particular ones certainly will do so.

4. CONCLUSION

Perhaps I have been too hard on Latour. His account is problematic, but it provides valuable resources for rethinking modernity, tradition, and sciences. From an influential base in science studies he does engage modernity as both a scientific and political issue; he is one of the most innovative and influential of the second (we could say) great generation of science studies scholars, and it is useful to be able to think along with one who is indeed engaged in some of the projects central to the present discussion. He provides compelling arguments against both modernist and postmodernist understandings of science and politics. He shows how science has appropriated political projects while anti-democratic politics has often been grounded in scientific legitimation of its own preferred public policy. He sees that because science and politics are so inextricably conjoined, they must together be transformed. He identifies the fact/value binary as a main culprit in this antidemocratic merging of science and politics. Distinctive to his critical treatment of this binary is his demonstration of its origins in one of the founding texts of Western civilization, namely Plato's Myth of the Cave. He shows how this myth gives illegitimate power, and much too much of it, precisely to scientists and their philosophies in the contemporary context. This is an illuminating treatment of this text. He is happy to abandon Plato, but not the Enlightenment, in the context of which he wants to redesign the fact/value distinction, or, rather, relocate its strengths onto another set of distinctions which avoid the limitations of the former.

Latour also critically analyzes another project of postpositivist science studies, the "social construction of science" accounts. He argues that these theorists both give too little attention to the role of nature in scientific claims and fail to take on the project of rethinking political philosophy and social theory beyond the call to extend their domains to the sciences. They substitute a misleading reification of "the social" for that of "the natural."

The pro-democratic social movements Latour criticizes have argued that there are two difficult tasks for dominant groups. First, they must learn to listen to these "Others," the better to grasp where they think partial common grounds can be found and constructed. Second, they must learn to live with residual elements of "difference," which can neither be denied nor appropriated into the dominant group's understanding. Indeed, they must learn to see some kinds of such differences as immensely valuable for science and for politics. According to Latour, such claims to difference constitute intolerable cognitive and moral chaos. Rather, I suggest that they result in an intolerable recognition on the part of the dominant social groups, in which Latour positions himself and his analyses, that the active cognitive and political agency of Others—persons who really are Other than us—is crucial to the success of one's own projects. Latour is wrong that the alternatives with which we are faced are limited to the modernist philosophy of science, Latour's proposal, or cognitive and political chaos. There are other possibilities produced by the pro-democratic social movements.

2.

THE INCOMPLETE FIRST

MODERNITY OF INDUSTRIAL SOCIETY

Beck

— — — —

The situation of the excluded . . . reflects back into the centre of modernity in industrial society and not just in the form of violence and loss of civility. It is reflected equally in the disruption or even destruction of the pretensions and foundations of institutions that live on the fiction of overcoming the problem of such "enclosed outsiders."
—BECK, *Reinvention of Politics*

THE SOCIOLOGIST ULRICH BECK is interested less in the classical Western philosophy of the modern which has been Bruno Latour's target. Beck's *Risk Society* (RS)[1] appeared in German in 1986 and, along with the work of Anthony Giddens (*Consequences of Modernity*) and others, opened up new ways of thinking about the dark side of modernization. More so than the other risk society theorists, Beck's work focuses on the relations of science to politics. It has had wide influence in Europe, though it is not much discussed in the U.S./European field of science studies. This is most likely because it does not produce the microsociologies, microethnographies, or microhistories of a particular science's lab or field work which, innovative and illuminating as these are, seem to have become the single most important requirement for recognition and engagement as this field has matured and is becoming institutionalized. Beck engages with Western science and its effects instead through macrosociological analyses.

Beck's analysis shares a number of perceptions and themes with Latour's account, as well as with those of Michael Gibbons, Helga Nowotny, and other contemporary science studies critics. He shows important limitations not only of modernity and modernization theories but also of postmodernisms. He provides a compelling argument for how science has appropriated decisions which belong in the realm of politics, while politics persistently uses scientific standards and facts prematurely to close down democratic discussion. In Beck's account one can begin to grasp that this kind of appropriation seems to signal the beginning of the end of what have been claimed to be some of the most distinctive features of modernity. Some of the major institutions of modernity, ones supposedly at least relatively autonomous, appear to be de-differentiating and reaggregating, as science seems to be rapidly losing its former relative autonomy from society. (This theme is pursued also by Nowotny et al. in a different way.) Contrary to Latour, Beck argues that we have indeed been modern. But we are getting less and less so. Is modernity shrinking or even disappearing?

Beck also recognizes that science and politics must together be transformed in pro-democratic ways. He, too, is optimistic about this project, though not without reservations. His own diagnosis and prescription are different from Latour's in significant respects. Yet these, too, suffer from inadequate engagement with feminist and postcolonial analyses, as well as with the integration of Western sciences into new forms of the global political economy. Nevertheless, his diagnoses and prescriptions provide useful resources for transformative projects that do not have such vulnerabilities.

In light of our interests here, what is distinctive about his work? Beck draws attention to how people increasingly perceive themselves to be endangered, and, in fact, *are* increasingly endangered, by modernization's scientific and technological projects and their products. He also identifies how modern sciences and technologies create conflicts of interest for themselves as they first create dangers, then diagnose and resolve them in ways that create yet further such dangers, which require more scientific diagnosis, and so forth. Sciences seem to have

become risk-profiteers. Thus where Latour's focus was on philosophies of modern sciences, their faulty ontologies, and the systematic
ignorance about the social/natural hybrids and networks of which our
daily lives are constituted, Beck's is on the production of inequality and
ecological destruction which industrial modernity, which he refers to
as the incomplete "first modernity," produces and legitimates, and on
the sciences' inadvertent interests, in the name of cognitive and social
progress, in creating an increasingly dangerous world. However, a
reflexive, second modernity is already beginning to emerge, he argues.
(In Beck's work, we shall see that the term "reflexive" refers to two
distinct phenomena, though Beck may not always recognize that he is
using the term in both ways.) It does so through the development of
kinds of epistemic and political agents and actions which "first modernity's" sciences and philosophies cannot even recognize as such.

Beck's social theory is much richer and more complex than Latour's,
one critically grounded in the history of modernization theory and in
extensive empirical studies of modernization and its problems. Our
focus here doesn't permit attention to the full range or complexity of
Beck's thinking. And his projects are still not finished since he is
producing new analyses yearly. Nevertheless, we can follow some central themes in his arguments which are relevant to the concerns here.
The sections below focus on how it is the successes of modern sciences
and technologies that themselves produce risk society, on the different
sources and forms of transformation of science and society already
under way, on how not only modernity itself but also attempts to transform it both produce countermodernities, and finally on several unfortunate limitations to Beck's account. He starts off with the recognition
that around the globe we now live in a world of terrifying risks.

1. MODERN SCIENCES AND TECHNOLOGIES
PRODUCE RISK SOCIETY

These days our everyday experience is of threats and risks. Now we live
in a context of generalized employment insecurity; the middle classes
as well as the working classes no longer can feel assured of continuous
employment or traditional retirement benefits. We fear the kind of economic crises that have sunk national economies virtually overnight,
wiping out what seemed secure futures and leaving the already least

advantaged economically and politically even more immiserated. We fear pandemics such as AIDS, SARS, or Asian Bird Flu, which can spread around the world with little warning, little possibility of immunity, and often few known remedies.

New kinds of environmental destruction seem daily to threaten our health, our lives, and the natural resources upon which human life depends. Damaging forms of radiation, whether from armaments, power plants, work sites, household products, or outer space, seem impossible to predict let alone to eliminate or control. "Everywhere pollutants and toxins laugh and play their tricks like devils in the Middle Ages. People are almost inescapably bound over to them. Breathing, eating, dwelling, wearing clothes—everything has been penetrated by them" (RS 73). The effects of genetically modified foods appear to be little known, yet it seems impossible to stop agribusiness from producing and selling them. There are also the dangers from second-hand smoke, mad cow disease, urban crime and violence, ever-increasing drug use, and a host of other contemporary phenomena—including, we can now add to Beck's list, biological, chemical, and military terrorism and, for the United States and its shifting military targets, engagement in seemingly permanent states of war. The already least advantaged groups globally have the fewest economic, social, and political resources to protect themselves against such dangers. Thus the conditions of risk society increase inequality globally and locally. Moreover, he argues, all of these dangers are enabled through modernity's sciences and technologies. Yet these were supposed to be the sparkling exemplars of rationality in modernity's crown of enlightened social progress.

Who is to blame? No one and everyone seems responsible for these risks. So no one is made accountable for them. It is impossible to prove with certainty—or, rather, what governments and their legal systems, let alone the risk-inducing industries, regard as a reasonable degree of certainty—that the toxic industry upstream, the tobacco industry, or agribusiness's use of pesticides and chemical fertilizers was responsible for a particular pattern of increased incidence of cancer or some other diseases. No one can show with certainty that any responsibility for urban violence or terrorism should be placed on the still expanding, profitable, and minimally regulated arms industry. No one is able to prove that the U.S. Army Corps of Engineers and the oil industry should be held responsible for the deterioration of the Louisiana and

Mississippi shores and levees in ways enabling the destruction of New Orleans and other Gulf cities during Hurricane Katrina. No one can show with certainty who is responsible for producing the greenhouse gases that cause global warming and consequent rising oceans, which threaten similarly to flood every low-lying shoreline city in the United States and around the world within the coming century or perhaps even sooner. Rather, supposedly scientific standards of proof are invoked by corporations, governments, and other institutions seeking to avoid responsibility for these disasters. Thus we all seem doomed to continue to experience these increased threats to life and health. We live in an environment of manufactured uncertainties and institutional irresponsibility. Modern sciences and technologies and their philosophies seem to be at the center of this situation.

Moreover, Beck argues that this risk society is not the consequence of failures of modern industrial societies but, instead, of their successes. It is the successes of science and engineering projects that have enabled scientists and engineers to convince themselves and us that they can predict and control nature. Consequently they think that they should have a monopoly on decisions about what constitute reasonable standards for prediction and control, as well as about which social considerations are and are not relevant to such processes. They and they alone have the expertise to make such decisions, they claim. Thus modern sciences have been permitted to appropriate as purely scientific and technical matters political decisions about how we shall live and die. Beck's focus is on the consequences of patterns of modernization which have been created by undeserved but widespread belief in the effectiveness of modern sciences and their philosophy.

Modern sciences and technologies are always already permeated by political projects, Beck argues. Indeed, nature itself ceases to be purely "natural" as it is absorbed into politically distinctive scientific and technical projects. (Recollect Latour's arguments about how nature, too, is an active agent—an "actant"—in our worlds.) "This *modernization of nature* produces the situation of being able to create 'nature' which is no longer that, and not just the image of nature" (*Risk Politics* [RP] 87). It is not just that new representations of nature are created by modern sciences and technologies. The latter also create forms of nature itself, which thereby become parts of social worlds with which humans interact.

Thus modern sciences, their epistemologies, and philosophies of science have permitted the development of only semi-modern societies, he proposes. We can understand Beck to be arguing that (the first) modernity's social categories, such as the natural vs. the social and the political, and experts vs. non-experts, have blocked our ability to understand its own processes. So, too, do its theories of the effects of scientific change, such as that scientific processes do not change the nature they study.

But this is only half of the story. The other half is about politics and the realm of "the political." In RP he explores the reinvention of politics, which is necessary if modern sciences and technologies are to be redirected toward decreasing social inequality and restoring the environmental resources upon which humans and all other life depend. He pursues into the realm of the political the question he asked in RS: "How can the risks and hazards systematically produced as part of modernization be prevented, minimized, dramatized or channeled?" (19). The semi-modernity which the most industrialized contemporary societies have achieved must itself be further modernized. We must complete the modernization projects which contemporary sciences and politics have only begun. Let us note at this point already that while Beck is interested in transforming modernity's social order in general, it is primarily the interactions between Western society and Western sciences which he keeps in focus.

Unfortunately, we cannot expect the sciences enthusiastically to support this project of completing the modernity program. It is problematic that contemporary sciences and their philosophies are not likely to look kindly on the kinds of transformations necessary to make possible fully modern societies and their more self-critical, reflexive sciences, Beck points out. They will resist the development of sciences which stand a chance of eliminating or even diminishing the extent and power of risks in society. The problem is that modern sciences profit too much from the existence of these risks. "Science is one of the causes, the medium of definition, and the source of solutions to risks, and by virtue of that very fact it opens up new markets of scientization for itself" (RS 155). The problem is not just that the actual and perceived threats of the risk society are created by modern sciences and their technologies. It also involves modern scientific and technological experts who are asked to analyze and measure these risks. Then the same experts are called upon to block and eliminate such risks. Finally, the solutions invariably, sooner or later, generate new risks, which the experts must again detect, identify, measure, resolve, and so forth. Why should the sciences happily embrace an end to this cycle from which they so richly benefit?

To summarize, Beck's point here is that risk society is knowledge-dependent. Modern sciences and technologies profit from their monopoly of the production of information ("truths") and thus on the continuation of the production of further risks which their epistemologies and philosophies of science made possible in the first place. Sciences in effect manufacture the risks that they then expertly analyze and expertly resolve, thereby manufacturing new rounds of risk. Decisions that will affect how we live are made on scientific and technological grounds that bypass the democratic processes to which political decisions are supposed to be subjected.

One must be clear about what is not being claimed here. Beck is not saying that individual scientists and engineers, or even science and technology institutions collectively, overtly or even covertly, *intend* that their research proliferate risk. To the contrary, I think Beck would agree that scientists and engineers, like most other people, usually see their work as that of pioneers, the advance guard of knowledge which will sooner or later produce social progress. Nor is he saying that it is the work of scientists and engineers *alone* which is responsible for the Risk Society. They need the partnership with other powerful social institutions that sponsor and fund scientific and technology research and regulate it to their own ends. The success of such partnerships is necessary for all parties to them.

We can see that this state of affairs protects scientific ignorance. An intricate conceptual apparatus protects individuals from having to understand—perhaps encourages them not even to question—the actual social consequences of their research or the function that this conceptual apparatus plays in protecting their systematic interested ignorance (cf. Tuana and Sullivan, "Epistemology and Ethics of Ignorance"). This problematic conceptual scheme includes rigidly maintained distinctions between the scientific and the social, between real science and its technologies and applications, between basic and mission-directed research, and between contexts of discovery and contexts of justification. It includes also belief in the necessary convergence of scientific and democratic processes, and thus in conceptions of scientific communities as "little democracies." Consequently, one reason scientists have difficulty appreciating the new, post-Kuhnian sociology, history, ethnography, and philosophies of science is that these new "sciences of science" reveal the lack of empirical support for this scientific self-image.[2] They also reveal the professionally self-interested reasons for maintaining this image, and the immense efforts necessary to do so in the face of increasingly widespread scepticism about its accuracy. The recent social studies of science and technology challenge scientists' authority about what science is and does, that is, about "the nature of science," as the science education discussions refer to the topic (Aikenhead, *Multicultural Sciences*). Most controversially, the sciences of science also challenge scientists' monopoly on authority about nature itself. They reveal how this apparatus has not kept scientific work autonomous from social and political influences, but has rather worked to maintain scientific and popular belief in such fictions.

A reader familiar with four decades of science studies can wonder if Beck, like many of his fans as well as some of his critics, has overemphasized the break between his first and second modernities in the sense that many of the features which he tends to present as new to second modernity were already fully visible to the field of science studies in the first modernity. For example, how autonomous was Western science ever from such macro social/political projects as the "voyages of discovery" and subsequent colonial enterprises, capitalist industrialism, and the various nationalisms and militarisms which necessarily accompanied such projects? Was science in fact disaggregated from pre-modern social and political institutions while, from the beginning, it was central to newly arising modern ones? What about the persistent male-supremacist gendering of modernity and its scientific institutions, practices, and cultures, to which we turn in chapters 4, 6, and 8? What do these histories say about modernity theories? Is it the self-representation of science, especially as elaborated by scientists and science policymakers in the 1950s, against which Beck's argument is most telling? To what extent is this false self-representation, elaborated into a supposedly universally valid philosophy of science, in historical fact a reaction to threats of even further intrusion into scientific and technological institutions of undesired social and political projects, or even of public oversight? (cf. Hollinger, *Science, Jews, and Secular Culture*). We turn to such issues below and in subsequent chapters.

To return to Beck's narrative, it is not just the sciences which will resist the kinds of changes necessary to "modernize modernization." It is also a problem that politicians and political institutions, including the disciplines of political philosophy and especially sociology (his own discipline), also will resist them (RP 32ff.).

> Highly specialized empirical social research presumes a constancy of categories and hence a high and actually rather rare social stability. . . . Western sociology presumes not only stable clientele relationships, but also a non-revolutionary social order constructed upon long-range—more precisely, industrial—certainties and antagonisms, which are "calculable" in the truest sense of the word and change according to probabilities. Collapses, the division of countries, the blurring of coordinates and the disappearance of entire groups of countries and military structures are not foreseen. . . . As a corrective measure it is necessary to "invent" and standardize methodologically a regulated opposition of social theory, social empiricism and social experience that will permit even external, extreme and explosive things to be incorporated within the horizon of that which is sociologically conceivable, observable and explicable. Otherwise, the horizon of sociology will actually shrink down to the mathematically expanded horizon of the nouveau-riche middle-class imagination, which excludes everything that transcends or threatens it: eruption, erosion, transformation, reformation. As long as such a methodology is lacking, hypotheses must be gathered and put together into empirically substantive theories. (RP 18–19)

Beck is arguing that social science disciplines, too, are highly invested politically and intellectually in the limited understandings of science and of the social and political which have maintained their dominance (cf. Heilbron, Magnusson, and Wittrock, *Rise of the Social Sciences*). Here he is talking about reflexivity both as strongly self-critical practices of sociology as a discipline but also as the unpredictable consequences of modern tendencies, the unpredictability of which must be accounted for in more completely modern thought than sociology currently can achieve—a point to which we return.

How are we to escape this dreadful scenario? In spite of opposition, we must proceed anyway and complete the modernity which industrial society only began. Thus, in contrast to both Latour and the postmodernists, who propose abandoning modernity, he conceptualizes the present moment as a break or shift in direction *within* modernity. Risk Society is the consequence of incomplete modernization. We have experienced only the first, industrial modernity, he argues. Only more strongly reflexive (self-critical) practices can open up scientific and technological decisions to appropriate democratic rule, including appreciation of their reflexive (unpredictable) consequences. We need the second, reflexive modernity which can provide more extensive scientific knowledge-seeking, broader rationality, greater objectivity, and a more robust appreciation of uncertainty.[3]

Thus, like Latour, Beck inveighs against the division of the world into nature, culture, and discourses. There is no pure nature or pure culture, let alone representations of either which have no social sources or consequences for nature and culture. Like Latour, Beck also rejects postmodernisms since they simply abandon the crises of the risk society. Moreover, they ignore the emergence of a new kind of modernity. "A new kind of capitalism, a new kind of economy, a new kind of global order, a new kind of society and a new kind of personal life are coming into being. . . . This is not 'postmodernity' but a second modernity" (WRS 2). This second modernity requires an expansion and reconfiguration of Enlightenment projects, not a rejection of them. "Where most postmodern theorists are critical of grand narratives, general theory and humanity, I remain committed to all of these, but in a new sense. To me the Enlightenment is *not* a historical notion and set of ideas but a process and dynamics where criticism, self-criticism, irony and humanity play a central role. . . . [M]y notion of 'second reflexive modernity' implies that we do not have *enough* reason . . . in a new postmodern meaning to live and act in a Global Age . . . of manufactured uncertainties" (WRS 152).

2. TRANSFORMATION IS ALREADY UNDER WAY

Beck is optimistic about the possibility of such a transformation since he sees it as already under way. He identifies three changes in the nature and practices of the production of scientific knowledge which contribute to such a transformation. In all three cases, the production of scientific knowledge has already begun to escape monopoly by the sciences themselves.

Sciences of science
One is the expansion of the scientific attitude and practices into the systematic study of sciences and technologies themselves: the development of "sciences of science" in the field of critical science and tech-

nology studies. In this way the production of scientific knowledge has escaped the monopoly of official laboratory and field scientists. Beck cites Latour's work as an example of this kind of critical science and technology studies (wRS 150–152). We need a truly reflexive modernity, he argues, one that can scientifically describe, understand, and explain its own principles and practices, which industrial modernity cannot. The sciences of science do this for one powerful modern institution. This project is not an idle exercise in simply turning the tables on scientists—doing unto science what science does to everything else in the world. Rather, the issue is that we need to be able to predict and explain how the institution of science gets so powerful and how these powers are illegitimately and without notice exercised in the political realm. Of course such sciences of science will themselves have to practice a critical reflexivity to bring into focus as much as possible their own assumptions and practices in the worlds in which they study the sciences. (And sciences of science presumedly also have unpredictable consequences—they, too, are reflexive in this second sense— which will have to be taken into account in their own analyses.)

Beck perceives that it is not only professionals in science studies who can help to generate this transformation of sciences and the political. Such workers can also be found in any and all scientific and technological work sites; they all can make important contributions to this project. Of course it is also the case, he points out, that "vocations and professions are (possible) foci of bourgeois anti-politics" (RP 157) and that they constitute "a centre of obstinacy for the self-assured individualist" (RP 157). Yet the "expert rationality" of these fields already has been troubled by changing conditions around and even within them: "An essential role is played here by the *issue of how deeply alternative activity affects and splits even the ranks of expert rationality.* Until now, this was unthinkable, or at least not a concrete threat. Three conditions have changed this: the transition from simple to reflexive scientization, the ecological issue and the penetration of feminist orientations into the various professions and fields of occupational activity" (RP 157; original italics). Moreover, Beck here points to the way the "stability of classical industrial society" is challenged by "feminist critiques of science and the professions, whenever they are not content with merely denouncing the professional exclusion of women, but go on to criticize the professional monopoly on rationality and praxis and to redefine and compose specialist competence with intra-professional acumen and methodology" (RP 158).[4] As such critiques disseminate through disciplines and professions, the first-modernity foundations of the authority of such fields become controversial, enabling scientific and political discussions which could not find space in the rationality of first-modernity. To put the issue another way, new, more democratic relations between authority and scientific speech seem to be emerging.

Conflicting experts

Second, Beck points out that the Risk Society forces every one of us to participate in the production of reliable empirical knowledge. We must do so because the successes of the sciences proliferate experts who continually disagree with each other. "Experts dump their contradictions and conflicts at the feet of the individual and leave him or her with the well-intentioned invitation to judge all of this critically on the basis of his or her own notions" (RS 137). Consequently, "science becomes more and more necessary, but at the same time, less and less sufficient for the socially binding definition of truth" (RS 156). Are smoking and second-hand smoke harmful or not? Do vitamins and herbal remedies improve our health and longevity? Which ones and how much should we take? Should we increase our intake of fish, since fish oils seem to provide protection against some cancers, or decrease the amount of fish in our diet since fish seem to pass on environmental toxins? Under what conditions should we take the radiation and chemotherapies recommended by Western biomedicine or the herbal remedies recommended by German pharmacologists to stop the growth of cancers? Will the benefits of genetically modified foods outweigh their risks? The proliferation of conflicting expert opinions on each of these topics requires that each of us must figure out how to make informed decisions about our own nutrition, health, and safety and that of our children and other dependents. We must collect a great variety of expert opinions, and then take an experimental attitude toward our chosen remedies. We are forced to become partners with scientists in the production of reliable scientific knowledge. The production of scientific knowledge escapes the monopoly of conventional scientist experts in this second way.

Questions for sciences from everyday experiences

Finally, everyday experience achieves much greater importance in knowledge production projects for escaping Risk Society. "In risk-

society civilization, everyday life is culturally blinded; the senses an-
nounce normalcy where—possibly—dangers lurk. To put it another
way, risks deepen the dependency on experts. A different way of han-
dling ambivalence thus presumes that *experience* is once again made
possible and justified in society—even against science. Science has
long ceased to be based on experience; it is much rather a science of
data, procedures and manufacturing" (RP 123).

Thus another kind of science is already emerging in "the civilization
of danger." Side by side sit "the old, flourishing laboratory science,
which penetrates and opens up the world mathematically and technically but is devoid of experience, and a *public discursivity of experience*
which brings objectives and means, constraints and methods controversially into view" (123). Each has its distinctive features, strengths,
and limitations:

> Laboratory science is systemically more or less blind to the consequences
> which accompany and threaten its successes. The public discussion—
> and illustration!—of dangers, on the other hand, is related to everyday
> life, drenched with experience and plays with cultural symbols. It is also
> media-dependent, manipulable, sometimes hysterical and in any case
> devoid of a laboratory, dependent in that sense upon research and argu-
> mentation, so that it needs science as an accompanist (classical task of
> the universities!). It is thus based more on a kind of science of *questions*
> than on one of answers. It can also subject objectives and norms to a
> public test in the purgatory of oppositional opinion, and in just this way it
> can stir up repressed doubts, which are chronically excluded by the
> blindness to threats and consequences of standard science. (RP 123–24)

Thus Beck brings into focus how these three kinds of expansion of
scientific processes are already well established and increasingly exercised. They contribute to a democratization of science in which pre-
viously inaccessible political processes are made observable and production of needed information is accelerated. Yet a fourth source of
transformative changes in the nature and practices of the sciences
themselves can be found in postcolonial and feminist foci in the sciences, as the chapters following will delineate. Beck refers to part of
this work above when he praises feminist work for criticizing the "professional monopoly on rationality and praxis" and for going on "to
redefine and compose specialist competence" (RP 158). Because sciences and politics so penetrate each other's supposedly autonomous

territories, the transformation of each requires changes in the other. Thus Beck also identifies important transformational potential for society, and by implication for the sciences, in new ways of organizing political activity.

Reflexivity

Beck's notion of reflexivity is central to his linkage of new sciences to new politics. He uses the term to characterize the emerging "second modernity." As I have already indicated, the term itself can be confusing since it is used with at least three different meanings in science studies and in contemporary intellectual life more generally. Beck links two of these meanings in a particularly fruitful way. He occasionally defines it as simply "self-critical"—science must become critical of itself. But elsewhere he makes clear that "reflexive modernization" is to refer to a particular kind of self-criticism, namely applying the principles of scientific skepticism to the very foundations and consequences of modernity's scientific projects. "Reflexive modernization here means that skepticism is extended to the foundations and hazards of scientific work and science is thus both generalized and demystified" (RS 14). Here we have an expansion of scientific rationality, applying to the foundations and consequences of our (Western) sciences themselves the principles that they apply to their conventional research objects. Modern industrial-era science has in effect forbidden this insofar as it refuses to assess scientifically its own grounds and consequences. "Reflexive modernization means not less but more modernity, a modernity radicalized *against* the paths and categories of the classical industrial setting" (RS 14).[5]

Yet there is another notion of "reflexivity" here which is different and goes beyond the self-critical. "The concept of 'reflexive modernization' is at the centre [of the reinvention of sociology necessary after the end of the Cold War]. This does indeed connect with the traditions of self-reflection and self-criticism in modernity, but implies something more and different, namely, as is to be shown, the momentous and unreflected basic state of affairs that industrial modernization in the highly developed countries is changing the overall conditions and foundations for industrial modernization. Modernization—no longer conceived of only in instrumentally rational and linear terms, but as refracted, as the rule of side-effects—is becoming the motor of social history" (RP 3). That is, industrial modernization produces unforeseeable effects which overrun modernity itself, changing its very nature. Central in such effects are all the new risks to life on the planet produced by the successes of modernization.[6] Another effect which was unforeseeable to modernity in its initial stages has been the emergence of rational criticisms of the very foundations of modern sciences themselves. Also unpredicted have been the "subpolitics"—politics outside the conventional domain of state and inter-state governance—which have contributed to such new skeptical projects. While unimaginable in the first modernity, these effects are in fact produced by it. Classical modernity produces effects it cannot imagine or find intelligible. Including "counter-modernities."

3. THE DARK SIDE OF PROGRESSIVE TRANSFORMATION: COUNTER-MODERNITIES

Beck alone of our three science studies theorists also brings into sharp focus the possible dark side of what comes after the "first modernity." In his view, we could help a reflexive, politically and scientifically progressive second modernity to flourish, or we could find ourselves stuck with counter-modernities, already all too visible, which block the most progressive tendencies of both the first and second forms of modernization (RP chap. 2).[7]

In the first place, modernity itself produces counter-modernities: "The concept of 'counter-modernity,' as the very name suggests, is conceived of as an antithesis to modernity. There would be nothing very surprising in this negation if this contradiction/connection were not conceived within modernity as an integral design principle of modernity itself. If modernity means questions, decisions and calculability, then counter-modernity means indubitability, indecision, incalculability and the attempt to force this indecision, contrary to all modernity, into a decision in modernity. . . . In the term 'counter-modernity,' the word 'modernity' thus also has an adjectival sense: modern counter modernity" (RP 62). Thus counter-modernity is a constructed certitude which is strengthened by modernity itself (RP 62). However, in the second place, the emergence of reflexive modernization "provokes counter-modernization, in all forms: religious movements, esotericism, violence, neonationalism, neo-racism but also the renaturalization of social relationships by genetic and human genetic trends. All of these live from the same promise to create new certitudes, or, better, rigidities, in one way or another, in order to put an end to permanent doubt and self-doubt" (RP 90).

Beck also notes that some observers have identified counter-modernities as fundamentally "a 'patriarchal protest movement in a great variety of forms' being articulated and forming into ranks in modern fundamentalism" (RP 68): "As Martin Riesebrodt (1993) shows in a comparative study of American Protestants early in this century and Iranian Shiites since the sixties, these fundamentalist tendencies need to defend not just their beliefs and readings of the holy scriptures, but their way of life as well, against effective competing interpretations and demands. In the dawning experience of relativity in modernity, faith becomes a matter of decisions and reasons, but this is certainly not a matter of visions of celestial justice. Rather, in view of the alternatives that are impinging—from women working outside the home, to the diminishing ability of parents to reach their children, all the way to the frequency of divorce and 'permissive' sexuality—the core of traditional ways of life is being put at stake" (RP 68). Beck shows the importance of recognizing that these counter-modernities thus are not residues of pre-modernity or "tradition," as both they and some of their critics claim. Rather, counter-modernity forces are an inevitable product of both industrial modernity's search for scientific and technological certitudes and reflexive modernity's doubts about such certitudes.[8]

Since reflexive modernity itself nourishes counter-modernities, he argues that the issue cannot be how to banish them. Rather, the question we must consider is "what kind of counter-modernity" do we prefer? (RP 90). For example, we can ask about the articulate enemies of environmentalism, "How can the counter-modernity of environmentalism be restrained, that is, how can its ascetic and dictatorial tendencies be connected to the extravagant tendencies and the freedom of modernity?" (RP 91). Similar questions must be asked of the counter-modernities of other subpolitics. Such questions are crucial for developing a transformed modernity which can avoid at least the worst of the scientific, technological, and political perils created by the first modernity. And they suggest the necessity of a different kind of revolutionary politics than those modeled on the political revolutions occurring in the eighteenth through the twentieth centuries, which have been idealized in different ways by both right and left political philosophers and activists. Beck's discussion of counter-modernities raises important questions for political philosophy as well as for democratic transformation activism.

There is much, much more that is interesting in Beck's work for those concerned with modernity and its sciences. But I hope there is here enough of a sense of main themes in his work to enable reflection on them. What are the limitations to this account?

4. LIMITATIONS

Critics have had a lot to say about the limitations of Beck's accounts.[9] In light of our purposes here, I will not take up the argument that he tends inappropriately to universalize the concerns and conditions to which the Green Movement in Germany responded. Nor will I pursue the complaint mentioned earlier, with which I agree, that he overemphasizes the break between industrial and reflexive (self-critical) modernity, characterizing the former more in terms of its self-image than its reality and, thus, drawing attention away from elements of reflexive modernity already active in the First Modernity. In *World Risk Society*, Beck directly responded to these and many of the other issues critics raised. His focus on political change in the later book meets at least some of the other charges critics have made. Here I focus on four interconnected issues of which only the first has received much attention by other critics: class issues and the global political economy, issues raised by postcolonial studies, gender issues, and the epistemic status of his own account.

(1) Class: The global political economy

In RS, Beck's account gave the impression that he took class to have lost its analytic power in thinking about how people's lives are shaped now and will be in the future (Draper, "Risk, Society, and Social Theory"; New, "Class Society or Risk Society?"). RP keeps class issues in better focus although, like many other thinkers, Beck vigorously distances his account from those that see class as the single motor of history. In what he admits is a simplistic description of the contemporary theories of modernity and modernization against which he proposes his own account, he makes clear that he attempts to avoid depending upon both of the main kinds of accounts of modernity. On the one hand are those who stay within simple modernization. These divide into "two intensely feuding schools—the *functionalists* and the *Marxists*" (22). The former focuses on post-industrialism and the latter on late capitalism. Both assume modernization must refer to industrial society processes alone.

On the other hand, the theorists of postmodernity also assume that modernity can refer only to industrial society processes. They simply dismiss principles of modernity. "Because modernity and industrial-society modernity are considered indissoluble, people jump straight from capitalist democratic industrial modernity into postmodernity, not a new modernity, when the historical falsity of the earlier model begins to dawn on them" (22). Both groups "thus rule out the subject of the present discussion, that is, the multiple modernities arriving in the wake of inherently dynamic modernization-as-usual, through the back door of side-effects (more precisely, slipping in behind the side-effects under the cover of ahistorical general ideas)" (RP 22).[10]

While valuing labor movements, and consistently marking how science and technology serve business and militarism and distribute their benefits and losses accordingly, he faults Marxian accounts for failing to perceive those inherent limitations of industrial modernity and its sciences which he identifies. He criticizes them also for lacking the resources to lead the way out of risk society. Whether it is the bourgeoisie or the proletariat that seeks empirical certitude and technological control from its sciences, those very achievements will return to imperil them and the nature upon which human flourishing depends.[11]

Yet Beck's attempts to block this kind of criticism do not succeed in protecting his account from a widespread perception that any compelling social theory or political proposal today must center the expansion of capitalist, economic, and political mechanisms as well as the kinds of economic, political, and cultural negotiations with such processes which are occurring around the globe. The interests of Westerners will often be at odds with the interests of groups less fortunately situated, and of groups with deeply different moral, political, and intellectual legacies and goals than the Enlightenment ones which uniquely get to count as progressive in those societies that have lived through industrial modernity. One place this problem reappears is in considerations of science and technology in global contexts, which Beck raises with only perfunctory gestures. We would hardly know from his account that Western sciences and technologies function in global political economies and that they are central resources for anti-democratic global social restructuring. To be sure, his thinking about *Risk Society* began before talk of globalization was so familiar. Yet it certainly has always been visible in post-World War II thinking about development policies and practices. At any rate, it certainly was fully visible by the end of the 1990s.

Moreover, there is no hint of how class relations within Europe and the West are replicated globally: that is, how the achievements of the West depend to a significant extent upon the immiseration of the "Third World," both at the periphery and in the industrialized nations. To be sure, he does point to the West's "constructed barbarianism," and as the opening quotation to this chapter also indicates, he certainly understands how powerfully what one could call the standpoint of the excluded brings into the spotlight constitutive failures of modernity and its institutions. Yet there is no account of the material social relations involved in such a construction. He does not intend to support the claim that industrial society class relations are the motor of contemporary history, but he also does not give much attention to the new kinds of class relations which have been produced in post-industrial societies in the North or globally. My point here is not to criticize Beck but, rather, to point out a limit to the usefulness of his work in this respect.

(2) Postcolonial studies

Beck's account lacks a necessary postcolonial studies focus on the history and present practices of Western sciences and global social relations; postcoloniality remains beyond the horizon of his project. Of course, Germany is not the United States. The end of the Cold War had a huge effect upon Germany, especially, so it is not surprising to find that clearly marked event compelling his thinking. And in Germany the term "racism" consistently brings up thoughts of the Jews and the Holocaust, and only secondarily and distantly the histories of peoples of African, Latin American, Native American, and Asian descent that are centered in U.S. thinking. The situation of the Southern "guest workers" in Germany has not yet created a crisis of national identity in the way that the persistence of racial inequalities and increases in immigration (and the consequent increase of citizens of non-European origin) do in the United States. (Though one can certainly understand that Germans had their full share of national identity crises in the last century.) Postcolonial science studies will be the topic of chapter 5; here I just mention a couple of its themes relevant to Beck's account.

First, he does not recognize that the development and modernization of the West were materially as well as ideologically *built on the exploitation*, de-development, and "constructed traditionalism" of the societies which European expansion encountered, from 1492 through the events in today's newspapers. While occasional references to Eurocentrism appear, imperialism and colonialism do not. To be sure, criticisms of continuing imperial and colonial power can be read out of or illuminated by many of his arguments. But these forces of history which have from the beginning deeply shaped the West's conceptions of modernity, its social theories and political philosophies, and its modernization practices, including its sciences and their philosophies, remain over the horizon in his account. This foreshortened horizon has moral and political consequences for Beck's own concerns. For example, as one commentator points out, "It may be that the desire of members of the advanced world for a cleaner world amounts to a demand that much of the world seeking development should desist. That demand is probably immoral, and is anyway not politically feasible—making the problems that face us deeper than Beck imagines" (Hall, "Review of *Risk Society*" 346).

This lacuna appears in a number of contexts. Two relevant here are the possibility—indeed, reality—of multiple global modernities, and his perception of who will be the subjects of history effecting the transformation of science and politics. In the first case, Beck provides a rich account of multiple modernities within Europe, but there is no hint that he recognizes that other modernities have emerged around the globe as other societies left feudalism and/or encountered European modernity (see Eisenstadt and Schluchter, *Early Modernities*; Eisenstadt, *Multiple Modernities*; Blaut, *Colonizer's Model of the World*). These other modernities also have contributions to make to Western understandings of the problems with the West's industrial modernity, and they offer resources for imagining and implementing second modernities. Moreover, they have been, are now, and will in the future produce significant subjects—speakers, agents—of second modernities, a point to which I return shortly.

(3) Still gender-blind?

It seems ungrateful to fault Beck for not sufficiently taking account of feminist work when he has so diligently credited the power of women's movements and feminist analyses, as indicated above, especially throughout RP. In this respect his account contrasts with those of Latour and Nowotny et al. Beck understands the power of women's agency to change the public sphere. He identifies, in a marvelous phrase, the "earthquake of the feminist revolution" which has split open so many social projects and social institutions (RP 114). He insists that it is women's collective agency which is so important. Thus, in the terms of historian Joan Kelly-Gadol ("Social Relations of the Sexes"), which we take up in chapters 4 and 6, he appears to see women as fully human. This apparently obvious fact is still vigorously resisted by "public sphere" institutions. Consequently, he appears to see that in principle it is necessary to consider women's status and powers when assessing standards for progress in history, and to consider gender as a crucial element in his theory of social change. Thus it would appear that Beck should get an A+ on a feminist ledger.

Yet there are significant lacks in Beck's account. He draws attention to the power of feminist analyses which are not content only with raising issues of exclusion, but "criticize the professional monopoly on rationality and praxis" and "redefine more adequate standards for expertise and good method" (RP 158). In this last phrase he identifies a distinctive contribution of feminist science critiques, though without telling us what such standards would be or letting us see how they have influenced his own account. It is not obvious that they have done so. Moreover, though the account does mention women and women's movements, the particular cultural ways of organizing masculinity which have shaped the history of science and society are not in view. Gender relations, and gender as an analytic category, are appreciated in the abstract, but they do not seem to inform these analyses. Consequently the account cannot recognize that every issue is a feminist issue. We are left to wonder how modernity, modernization, and Third World development interact with preexisting gender relations and/or produce new ones of their own, and how gendered social relations interact with racial and imperial ones; how is gender relevant to the two notions of reflexivity?

He does make the valuable point that counter-modernities seem consistently to be patriarchal projects, but what about the first, industrial modernity? Was it, too, a patriarchal project? How are the drive for prediction and control, for value-free objectivity, for narrow instrumental rationality, for such a limited notion of "real science" or "good method" distinctively masculine projects? How is the second modernity to be an improvement for women over the first? Will the "inter-nalization" of gender relations, considered only externalities in theories of simple modernity, produce yet further institutionalizations of patriarchy?

Women's conditions have rarely improved as a side-effect of other changes since men do not want to see their gender privileges slip away. They are often willing to make great sacrifices to retain them (cf. Hartmann, "Unhappy Marriage of Marxism and Feminism"). The improvement of women's conditions has tended to require direct attention. Beck wants to include such concerns in the improved modernity he envisions, but we do not get to hear how women will fare in that vision. Moreover, in Beck's account the voices of women themselves are not heard, nor is there a plan to take advantage of their particular resources—of their standpoint on the constitution and functioning of the dominant institutions—when planning to transform science and society.

It must be noted that Beck and Beck-Gersheim (*Normal Chaos of Love*) undertook a fascinating and illuminating analysis of contemporary intimacy relations—a topic which feminism introduced to sociology. They show how what appear as "private troubles" in contemporary intimacy relations in fact are the appearance in everyday life of "public troubles" in relations between second modernity's work patterns and family relations in the context of the successes of women's movements. Here Beck and Beck-Gersheim are charting recently chaotic relations between two of modernity's great institutions—the economy and the family. I am going to ask a question which is no doubt unfair: what is the relation between this particular chaos and the worlds of science and politics in "Risk Society" on which Beck has focused in so much of the rest of his work? This question will no doubt seem to risk theorists absurd and, at any rate, impossible to answer. Yet we will begin to see a path to some answers to it in chapters 8 and 9.

Beck needs standpoint methodology, as it has been articulated in feminist accounts, to identify and explain the potentially positive effects not only of the "perspectivism," which he values, but of the resources provided by the social position of the exploited in any particular context (see Harding, *Feminist Standpoint Theory Reader*). Beck's references to the new skeptical analyses of the foundations and consequences of modern sciences, analyses grounded in an experiential "science of questions," come close to a central thesis of standpoint methodologies, but he does not quite grasp it. Standpoint methodology details exactly how everyday experience of exploited groups reveals the material and "conceptual practices of power" (Smith, *Conceptual Practices of Power*). Beck glimpses the power of thinking from the "situation of the excluded," but then loses this insight in developing his project. Indeed, the very language of inclusion leads in the wrong direction since it suggests there is nothing wrong with the way things are going except that the excluded happen not to be part of these processes. This language leads away from the idea that the excluded might have very different needs and desires which require the already included to change radically their existing ongoing projects. Starting with women's lives can reveal the grip of patriarchy on dominant conceptual frameworks and everyday practices, including those of modernity, modernization theory, and perhaps even Beck's own proposals. Starting off from the lives of the victims of development modernization policy can reveal how modernization theories and practices remain under the control of patriarchal, capitalist, colonial, and imperial conceptual frameworks (cf. Visvanathan et al., *Women, Gender, and Development Reader*).

(4) Status of his own account?

Let us close this section by focusing on the status of Beck's account.

On the one hand he argues for doubt, for the willingness to accept that one is probably wrong. He sees the cultivation of doubt as a gateway to new politics and new identities. "Perhaps doubt, mine and yours, that is, will create space for others, and in the development of the others, for me and us? Could this utopia of a questioning and supporting doubt form a basis, a fundamental idea for an ethics of a post-industrial and radically modern identity and social contract?" (RP 162). He promotes important "open" strategies (RP 43, 123): He values how people have learned to use their experience to "talk back" to power, even to its sciences; "sciences of questions" play a valuable complementary role to "sciences of answers" (58). He insists that the future is not foreseeable; that we cannot now identify the solutions to many of the problems facing us today (109). What he wants to create is a sense of possibility; things could be different (chap. 6, 168–69). These linked themes about the importance of doubt and the unforseeableness of the future provide a powerful bulwark against the search for certitude and "being right" which has characterized the history of expertise which he criticizes. He clearly understands the necessary and valuable "partialness" of knowledge claims, a point which has been central to feminist accounts.[12]

Moreover, he continually interrogates the conceptual frameworks of his own discipline which have shaped how most sociologists think about science and politics today (e.g., RP 18, 19, 21). He is clear that sociology itself serves industrial society's rulers (112). Sociology refuses to consider alternative modernities (175). Sociology serves power and it needs reformation. The great nineteenth-century modernization theorists were wrong.

Yet we must still ask at least one kind of critical question about the status of his account. Beck seems to think that it is Northerners who will create the new world for which he calls, and perhaps even only non-feminist Northerners, since there are no citations to specific feminist writings from which he has learned. "We doubt, and therefore there are many modernities and everything is starting from scratch!" (RP 162). But of course non-Western cultures do not follow this logic. They have no intention of "starting from scratch" once modernity is no longer identified only with the classical kind of Western modernization. Rather, they can now recognize their own distinctive forms of modernity and start from what they find most valuable in their own cultural legacies to construct societies and knowledge projects which affirm their histories of struggle and achievement. Feminists want to start from women's worlds to think about how to redesign modernization. And why should Westerners imagine that we should or can cast off our entire Western tradition?

Of course, Beck does not in fact intend to cast it off, for he carefully preserves valuable parts of the Western legacy (even though they no doubt are important parts of other cultural legacies also, a point he does not recognize), such as the important role of doubt and skepticism in the constitution of modern Western sciences—in imagining possibilities for the future. He wants to preserve the Enlightenment as a radical tendency, and even to complete (Western) modernity, not to banish it. Moreover, even if we did want to abandon the Western legacy, why would we want or be forced to start from scratch in a world with hundreds of rich non-Western cultural traditions? Why can't we learn from those non-Western agents of history whose actions do and will affect the West? Haven't we in fact always done so? (cf. Harding, *Is Science Multicultural?*; and Hobson, *Eastern Origins of Western Civilization*).

In fact, as Hall notes above, it is a continuation of the Western megalomania which Beck otherwise criticizes, to imagine that the West will constitute its future all alone. Beck's account needs the central presence of people besides Western bourgeois men as powerful and progressive agents of history and knowledge.

5. VALUABLE RESOURCES

Yet in spite of these limitations, Beck's account provides valuable resources for our project. Mentioned at the outset were those shared with Latour and Nowotny et al.: He does interrogate specifically the modernity of modern science. He shows how science and politics are so deeply imbedded in each other's projects that to transform one of them we must transform the other—a task few science studies scholars or political philosophers have taken on. He rejects postmodernism as a solution to this problem. He is optimistic about the possibilities of such transformation.

Additionally, we saw that five more strengths were distinctive to Beck's analysis. First, Beck has a much richer social theory than Latour, and he takes up important issues that Nowotny et al.'s also rich account does not. He invokes a more complex analysis of the social aspects of the production of scientific knowledge, and innovatively brings into focus some of the social effects of science's successes. Consequently he can direct readers to a richer proposal for resolving central problems with modern sciences and their philosophies. Second, he proposes not that we abandon modernity and its projects, but rather that we complete them. This is to be achieved through an expansion of the gem in the crown of the Enlightenment, one could say, namely, scientific rationality. One important way it is expanded is through a vigorous double-track reflexivity program which both questions the foundations and consequences of (Western) sciences' successes and insists on taking far more seriously the unpredictability of human actions and their consequences including, especially, modern sciences.

Third, Beck identifies several ways in which the sciences have also lost their monopoly on the production of scientific knowledge. Thus scientific rationality expands past its own borders in industrial modernity. Moreover, fourth, much of this expansion of scientific rationality arrives through new social movements, Beck argues. Of course there are only two of them that loom in his field of vision—environmental and feminist movements. But that is two more than either Latour or Nowotny et al. can see.[13] For Beck, these kinds of social movements are a crucial resource for transforming science and politics. My argument is that they constitute a fourth process through which scientists no longer monopolize the production of scientific knowledge: social movements also create distinctive expertise and the ability to shape knowledge projects.

Finally, while Latour and Nowotny et al. notice resistance to the kinds of transformations of science and society for which they call, only Beck recognizes how both modernity and contemporary transformations of it themselves produce these counter-modernities. Counter-modernities are not a residue of tradition, but a dynamic at the very heart of modernity and its transformations—past, present, and future. They can have most unpleasant articulations and effects; yet they also continuously invigorate social movements within the modern and new transformational projects. Consequently he poses a provocative set of questions to be asked about these tendencies which cannot ever be defeated or exiled, but instead must be recruited to less harmful interactions with modernity and its transformations.

Latour and Beck have raised provocative questions about the failed dreams of modernity and of the sciences which were to be so central to their production of social progress. A very different kind of study, this time of the new forms of producing scientific and technological knowledge which have emerged in Europe and the United States after the end of the Cold War, provides yet further resources for our project of rethinking modernity. We turn to this in the next chapter.

3.

CO-EVOLVING SCIENCE AND SOCIETY

Gibbons, Nowotny, and Scott

— — — —

Science has spoken, with growing urgency and conviction, to society for more than half a millennium. Not only has it determined technical processes, economic systems and social structures, it has also shaped our everyday experience of the world, our conscious thoughts and even our unconscious feelings. Science and modernity have become inseparable. In the past half-century society has begun to speak back to science, with equal urgency and conviction. Science has become so pervasive, seemingly so central to the generation of wealth and well-being, that the production of knowledge has become, even more than in the past, a social activity, both highly distributed and radically reflexive. Science has had to come to terms with the consequences of its own success, both potentialities and limitations.—NOWOTNY, SCOTT, AND GIBBONS, *Re-Thinking Science*

THE EUROPEAN SOCIOLOGISTS Michael Gibbons, Helga Nowotny, and Peter Scott have together and with other colleagues set out to document and analyze the distinctive forms that Western scientific research has taken since World War II and especially since 1989, which marked the end of the Cold War. That date is significant since federal funding for science and technology research in European and North American countries was sharply cut back at that point. This had grave effects on university science research and training, both of which had exponentially increased since the mid-twentieth century. No longer able to find research and teaching opportunities in university science departments, graduates flowed into research jobs in industry and in federal government labs. In these contexts, the production of scientific knowledge has taken new forms. It has had surprising consequences for the assumptions of philosophies of science (Gibbons et al., *New Production of Knowledge* [NPK]; Nowotny et al., *Re-Thinking Science* [RTS]).[1]

GNS use this shift in the forms of producing scientific knowledge to argue against the positivist-influenced philosophies of science which still prevail in science departments, in the media, and in popular thought. But they intend far more than just one more science studies criticism of positivist philosophies of science and epistemologies. They are interested in the interactive effects of such changes in the production of knowledge with simultaneous and partly independent evolving transformations in the social order. Thus they insist on the necessity of thinking simultaneously about the production of scientific knowledge and of "the social." The two—"Mode-2" society and "Mode-2 science" —evolve as a dynamic system, each helping to constitute the other.[2] In their early work they argued that the effect of such changes, on balance, will be to increase social inequality. In their more recent writings they still tend toward such an assessment, but propose that if the ongoing relation between sciences and their social worlds are properly understood, one can and should begin to identify ways to recommit sciences to advancing social equality. Moreover, such a recommitment in fact produces more desirable knowledge that is not only reliable by conventional standards, but also "socially robust," which amounts to a kind of super-reliability.

Their work is located at the intersection of science studies and science policy studies. The policy concerns mark their work in distinctive ways and set their analyses apart from those of Latour and Beck. For example, in their account it is "the social" rather than "the political" that is their focus. They do introduce interesting observations and proposals about the current organization of social life, yet the particular way this is conceptualized locates their work within a problematic lineage of management studies. However, their analyses do share themes with Latour's and Beck's work, as indicated in my introduction. They draw attention to the emergence of modernity's unpredicted "Risk Society," drawing here on Beck. They are specifically concerned with challenges to modern philosophy of science and to moderniza-tion practices which have been created by the new ways of organizing the production of scientific work. They have grasped some of the cen-tral themes of the arguments that the pro-democratic social move-ments have put forth. In fact their analyses, like those of Latour and Beck, provide powerful justifications for feminist and postcolonial projects even though they do not put their arguments to this goal. In turn, those social movements could effectively make use of some of their arguments.

Though their account makes for frustrating reading at several points in ways to be identified below, these two books are nevertheless im-mensely rich and raise issues well worth pondering. This work has received exuberant praise from some readers and a deeply skeptical reception from others. In *Nature* the distinguished French professor of technology and society, Jean-Jacques Salomon, says of RTS: "The topic of science as a social institution and its relationship with society has not been covered in such an original fashion since the first seminal papers by Robert Merton in the 1930s and Thomas Kuhn in the 1960s" (585). On the other hand, critics complained about "the book's un-stated obsession with questions of science policy and governmentality, and its implicit participation in building a new public image for sci-ence and society" (Linder and Spear, "Rethinking Science" 256); one also noted that "the authors focus on elite scientists and institutions," which has the consequence that the larger issue "of having better science in a better society is then backgrounded" (Rip, "Reflections" 321). Clearly a work which produces such a range of responses is raising issues about science and society that matter deeply to different groups for different reasons. I cannot possibly do justice here to the complexity of their claims or the interesting evidence they marshal for them. They make clear in the book that they want their "open, dynamic framework for re-thinking science" to be understood as "based upon four conceptual pillars: the nature of Mode-2 society; the contextual-ization of knowledge in a new public space, called the *agora;* the devel-opment of conditions for the production of socially robust knowledge; and the emergence of socially distributed expertise. Our conclusion, briefly stated, is that the closer interaction of science and society signals the emergence of a new kind of science: contextualized, or context-sensitive, science" (RTS vii). In the too-brief space allotted to them here, we can at least get a good sense of how they understand these four shifts, the ways these analyses support and even (unintentionally) call for the feminist and postcolonial projects, and then some serious problems in their account for which we can begin to identify solutions.

1. CHANGING CONDITIONS IN THE PRODUCTION OF SCIENTIFIC KNOWLEDGE: THE EMERGENCE OF MODE 2

This work began with a study commissioned by FRN, the Swedish Council for Research and Planning; it was intended as a resource for Swedish science policy. NPK was the result of a three-year study conducted in 1990–93 by Gibbons, Scott, Nowotny, and three of their colleagues. The study was carried out mostly in Europe, but the authors propose that it is equally relevant to science research in the United States. Moreover they suggest that social science research, too, has been affected by the changes they identify, though their focus is mostly on the natural sciences. Their concern is the kind of research—"Mode 2"—that has been replacing the "Big Science" model, which they identify as the apogee of "Mode 1." Mode 1 has been around for many decades or perhaps even centuries; it names the solidification of Newtonian scientific processes for earlier periods, and as these have been understood by conventional philosophies of science. Big Science research emerged after World War II (which was also the moment modernization theory reached its apogee—a point to which we return in later chapters). It featured complex, expensive projects employing large numbers of scientists who were disciplinarily trained and organized hierarchically. GNS are especially interested in the challenges which the subsequent production processes of Mode-2 science are presenting to universities and their disciplinary structures in which scientists have traditionally been trained and in which they have conducted most scientific research. They argue that this Mode-2 kind of organization of science training and research left universities increasingly unable to provide the highest-quality science education for the new post-Big Science forms of research.[3] In RTS they return to what they recognize was their neglect in the earlier book of analysis of the new forms of "the social" and how it co-evolves along with Mode-2 science.[4]

A number of changing social conditions shaped the shift to Mode-2 science and society. Two crises in particular destroyed confidence in modernization's ability to control either its sciences or its social worlds. The 1973–75 oil crisis was unpredicted. It, along with the emergence of the Internet, produced new sources and patterns of the production of scientific and technical knowledge through globalization and internationalization (RTS 6–7). Then the 1989 collapse of Eastern Europe and the end of the Cold War demonstrated the unpredictability of politics (RTS 8–9). It, too, had effects on modern science and technology since the latter's era of high funding had been tied to Cold War needs. Modernization seemed unable to control or even predict either its scientific or its political projects.

After 1989 federal funding was withdrawn from Cold War scientific research projects. But the universities still had in their production line the Cold War level of discipline-trained scientists. The output of these scientists became increasingly greater than could be absorbed by university labs themselves (RTS 73). This led to a migration of highly trained scientists, with their distinctive professional concerns and practices, into corporate and government research projects. On the receiving end, this migration was welcomed because of the new need for more scientifically trained researchers and managers in corporate research and development in the emerging "information society." Such a need was created by the new electronic organization of both manufacturing and the daily operation of corporations and government agencies in increasingly fast-paced global contexts. Corporations and government agencies were glad to have access to more scientifically trained researchers than pre-Internet and pre-cellphone production and management had required.

However, scientific research was organized differently in industry and government than in university laboratories, though it changed significantly as the university-trained scientists entered such research contexts. Several differences from university settings were particularly significant. First, in their new job sites, the scientists formed multidisciplinary research teams in contrast to the then still predominant pattern in university labs of research co-organized with disciplinary colleagues. Second, these teams tended to have a shifting membership as different skills and abilities were needed at different stages of each project. Scientists and engineers with particular skills, knowledge, and experience joined research projects only during the stages where their specialized expertise was needed. Third, these teams were focused from the very beginning of a research project on solving a practical problem. Thus the possible application of the research was envisioned from the beginning of the research project, in contrast to the situation for "pure research," which had long been articulated as the goal and the achievement of university-based research.

Meanwhile, corporate and governmental scientific projects and management styles increasingly were introduced to university labs. Simultaneously, elements of university management styles and "ways of life" entered corporate and governmental projects. For example, such university features as research seminars, less hierarchical organization of research projects, and the university look of campuses began to appear in corporate and governmental lab contexts. Thus, the decrease in federal funding of science had the unexpected effect of deeply undermining university scientists' monopoly on both the direction and the character of the production of scientific knowledge, as NPK points out. This is one force toward the social distribution of scientific expertise, about which we will hear more shortly. The exercise of scientific expertise was disseminated to and began to be produced in all kinds of private research projects, federal agency labs, corporate labs, and research councils. In RTS, these changes are conceptualized in this way: "The emergence of more open systems of knowledge production —Mode-2 science—and the growth of complexity and uncertainty in society—Mode-2 society—are phenomena linked in a co-evolutionary process . . . [T]he links between them can be understood in two different ways: in terms of the erosion of modernity's stable categorizations —states, markets and cultures—and also in terms of the transgressive, distributing effect of co-evolutionary processes. . . . Both interpretations tend to the same conclusion: science and society have both become transgressive; that is, each has invaded the other's domain, and the lines demarcating the one from the other have all but disappeared" (RTS 245). In Mode-2 science, the separation of science from application, and research from development, became difficult to detect. The boundary between "pure science" or "basic research," on the one hand, and mission-directed research, on the other hand became increasingly difficult to locate.[5]

So Modes 1 and 2 are both modes of modern science and its society. The focus here is on a shift in modernity. Mode 1 is defined as "the complex of ideas, methods, values and norms that has grown up to control the diffusion of the Newtonian model of science to more and more fields of enquiry and ensure its compliance with what is considered sound scientific practice" (NPK 167). In contrast, Mode 2 designates "knowledge production carried out in the context of application and marked by its: transdisciplinarity; heterogeneity; organizational heterarchy and transience; social accountability and reflexivity; and quality control which emphasizes context- and use-dependence. These features result from the parallel expansion of knowledge producers and of knowledge users in society" (NPK 167). We return shortly to these characteristics of Mode 2. Mode 1 exists alongside the new Mode-2 ways of producing knowledge; it probably will not be replaced by the latter, they suggest. Indeed, the old and new have ongoing interactions with and effects on each other.

2. THE CONTEXTUALIZATION OF SCIENTIFIC RESEARCH

"Science is always now contextualized," GNS argue. What does this mean? In twentieth-century philosophy of science, the life of a scientific project was "rationally reconstructed" as divided into two discrete stages: the context of discovery and the context of justification. The former was supposed to be the hallowed domain of scientific creativity. This was where particular situations in the natural world were identified as problematic, possibly explanatory concepts and hypotheses were identified, and a research design proposed which could test a hypothesis against observations of nature and against explanatorily competing hypotheses. Scientific problems could come from anywhere, it was said—from sun worship (e.g., Kepler), reflection on unpredicted events or processes (e.g., Curie), talking to the most knowledgeable scientists of the day, or, for all the philosophers cared, gazing into crystal balls. It was only when such hypothesis trials with their methodological controls had been designed that the context of justification was entered. Scientists' accountability and responsibility were to be called on only in the methodological processes of justification, and even these were understood in exceedingly narrow terms. Most importantly for the argument here, justification methods were to be "decontextualized," that is, detached as much as possible from social aspects of the worlds from which they had arisen and in which they were practiced.

Thus the two stages of research were to be kept distinct and discrete.

The context of discovery was to give free reign to scientists' creativity. Attempts to exert any kinds of controls over which problems got to the starting point of justification have been to this day treated as a vile and destructive intrusion into the creativity of scientists and the whole ethos of science. (Never mind national, military, or industrial priorities and funding, with their frequently racist, sexist, profiteering, and imperial research programs!) Only in the context of justification would the "bold conjectures" which had emerged from such creativity be subjected to the attempts at "severe refutation" which methodologically controlled trials could enact. Such refutation was thought to depend upon the purity, the value-freedom, of the justification process.[6]

RTS argues that in the new forms of research, discovery and justification are no longer discrete. Science is no longer segregated from society, nor should segregation be seen as desirable. Moreover, to the contexts of discovery and justification should be added two additional contexts which shape the results of research. Research is carried out from the beginning in, first of all, the "context of application" since Mode-2 research is already "mission directed." Thus this is one way in which society now "speaks back to science" through its major knowledge-producing social institutions such as industrial laboratories, government research establishments, research councils, and universities. These social institutions are often subjected to demands from kinds of social groups that expect research to benefit them, too. Thus the range of "stakeholders" which scientific projects must consult has been expanded, they argue. "In all these arenas the articulation of social demands and the expectations that research should yield socioeconomic benefits have become pervasive. To the extent that researchers and research institutions responded to such differentiated social demand, they were moving beyond Mode-1 science into the wider terrain of Mode-2 knowledge production in the context of an emerging, and co-evolving, Mode-2 society" (RTS 96).

Moreover, science must also function in the "context of implication," where scientific research has unforeseen and often unforeseeable consequences. "The process of 'reverse communication' is transforming science, and this, in the simplest terms, is what is meant by contextualization. . . . [C]ontextualization involves not just an increase in the number of participants, their institutional or disciplinary affiliations, their experience, interests and networks, or even an expansion of the lines of communication between them. It also evolves in ways . . . in which the shared definition of problems, the setting of research priorities and even to some extent the emergence of new criteria of what it means to do good science may be affected" (RTS 246). Most importantly, RTS says, contextualization pays attention to how "people" enter science, "as users, as target groups in markets or addressees of policies, even as 'causes' for further problems to be tackled, or as 'real' people in innumerable interactions and communicative processes, ranging from new modes of investment and financing for research, to legal regulations and constraints that shape the research process, to markets and media, households and Internet users, other scientists and millions of sophisticated and highly educated lay people" (RTS 246).

Mode-1 philosophies of science would see as a serious threat to the quality of research all this permeation of research projects by particular historical social contexts and by people with distinctive social/economic needs and demands. But Nowotny et al. come to a different conclusion. In Mode-2 science and society, such "transgression" increases the reliability of scientific knowledge, they argue. It makes scientific knowledge more "socially robust."

3. THE PRODUCTION OF SOCIALLY ROBUST KNOWLEDGE

Mode-1 science projects aimed for reliability. Achieving reliability required the de-contextualization of knowledge through the segregation of science from polluting social influences. In Mode-2 science, reliability is not enough. In fact the reliability of conventional "reliable knowledge" has been criticized as producing far too narrow understandings of nature and social relations. "Reliable knowledge remains the indispensable 'condition sine qua non' of the fact that 'science works.' But if reliable knowledge has been undermined, it is possible that this has occurred as much by the narrow reductionism of scientific practice as by any attempt to widen the range of stake-holders or more systematically to articulate the context in which science is produced" (RTS 246). In Mode 1 knowledge was incomplete because it would always be replaced by superior knowledge. But in Mode 2, knowledge is incomplete in an additional way. It is incomplete in the "sense that it is sharply contested and no longer entirely within the control of scientific peers. This shift involves renegotiating and reinterpreting boundaries that have been dramatically extended, so that science cannot be validated as reliable by conventional discipline-bound norms; while becoming robust, it must be sensitive to a much wider range of social implications" (RTS 246). So the kind of closure of scientific inquiry which Mode-1 projects could expect to reach is unavailable in Mode 2. In effect, scientific experiments never end since their effects and meanings remain perpetually open to reinterpretation and renegotiation by different social groups (cf. Galison, *How Experiments End*). Indeed, the epistemological core of science, which in Mode 1 guaranteed a reasonable assessment that closure of a particular scientific project had been reached, has turned out to be empty in the sense that it "cannot readily be reduced to a single generic methodology or, more broadly, to privileged cultures of scientific inquiry" (RTS 247). That epistemological core has dissolved and seeped away. An alternative way to put the point is "that the epistemological core of science is actually crowded with a variety of norms and practices" (RTS 246–47). That is, they are saying, it doesn't matter if we think of it as either empty or crowded in this way. We will return to this point.

This has a significant consequence. "The sites of problem formulation and negotiation of solutions move from their previous institutional domains in government, industry and universities into the *agora*. The *agora* is the public space in which 'science meets the public,' and in which the public 'speaks back' to science. It is the domain (in fact many domains) in which contextualization occurs and in which socially robust knowledge is continually subjected to testing while in the process it is becoming more robust" (RTS 247). So scientists, as this group was conceptualized in Mode 1, no longer have a monopoly on the production of scientific knowledge. Rather, "although we are not all scientists yet, many more of us are" (RTS 184–85). RTS here echoes Beck's point. Socially robust scientific knowledge is produced only through the social distribution of scientific expertise. How do GNS conceptualize this?

4. SOCIALLY DISTRIBUTED SCIENTIFIC EXPERTISE

One result of the emergence of the *agora* is that "the role of scientific and technical expertise is changing as expertise becomes socially distributed" (RTS 247). Contributing to this distribution in Mode 2 have been two phenomena. "The first is that two generations of mass higher education have increased significantly the proportion of knowledgeable social actors; there are now many more scientifically trained politicians and civil servants, industrialists and business people, who can no longer be treated as incompetent outsiders. The second is that . . . the pool of potential peers has been systematically diluted, [and] so eroding the coherence of generic scientific communities which can be distinguished from broader social coalitions" (RTS 49). Thus, RTS argues, "science has penetrated, and been penetrated by, society. It is in this sense that it is possible to speak of co-evolution." This social distribution fragments "established linkages between expertise and established institutional structures whether of government, industry or the professions" (RTS 247).

And the social distribution of expertise requires new narratives of expertise. These narratives are transgressive, collective, and self-authorizing.

> Narratives of expertise are constructed to deal with the complexity and uncertainty generated by this fragmentation. They have three characteristics. . . . They are transgressive in the sense that experts must respond to issues and questions which are never only scientific and technical ones, and in the sense that they must address audiences which never consist only of other experts. They are collective in the sense that the limits of competence of the individual expert call for the involvement of a wide base of expertise which has to be carefully orchestrated if it is to speak in unison. Finally, the social distribution of expertise mirrors the social distribution of scientific authority. Since expertise has to bring together knowledge which is itself distributed, contextualized and heterogeneous, the authority of expertise cannot arise at one specific site, or from the views of one scientific discipline or group of highly respected researchers, but precisely from bringing together the many different and heterogeneous practice-related knowledge dimensions that are involved. The specific "scientific" authority of expertise resides in the links that bind it together in its highly distributed form. In this sense, scientific and technical expertise is self-authorizing under conditions of Mode-2 society. (RTS 247–48)

Yet Mode-2 modernity seems in several respects to cast only a withered shadow alongside its buff Mode-1 ancestor and still present companion.

5. SHRINKING MODERNITY?

Modern philosophies of science and modernization theories are both clearly the target of RTS's analysis. "Science and modernity have become inseparable" (RTS 1). Science and technology have become "irreversibly identified with the modernization project" (RTS 180). RTS, like Beck, sees today the closure of a long period in the history of modernity (RTS 191). This period begins with, say, Newton; it reaches its apogee from the end of World War II until the end of the twentieth century. During this period modernization programs were widely presumed to provide compelling evidence for the soundness of modern philosophies of science and of modernity as a social program, RTS argues. Yet today, modernity seems to be dissipating; modernization seems to be departing from modernity; and scientists no longer have the monopoly of the production of scientific knowledge which seemed integral to modernity.

Dissipation

Differentiation of social institutions is always taken to be a defining mark of modernity. Yet RTS points out how the segregation from each other of the great institutions of modernity in fact is decreasing as they transgress and become porous to each other (RTS 48). This change is accompanied by a similar problem with the "great binary categories" which also defined modernity. "The great conceptual, and organizational, categories of the modern world—state, market, culture, science—have become highly permeable, even transgressive. They are ceasing to be recognizably distinct domains. As a result, common-sense distinctions between the 'internal' and the 'external' are becoming increasingly problematic, a change which has radical implications for demarcations between science and non-science and for notions of professional identity and scientific expertise. . . . [J]ust as the boundaries between state, market, culture and science are becoming increasingly fuzzy, so too are those between universities, research councils, government research establishments, industrial R&D, and other knowledge institutions (for example, in the mass media and the wider 'cultural industries')" (RTS 166). Moreover, there are additional ways in which the foundations of Mode-1 modernity are under siege.

Modernization departs from modernity

In this dissipation of modernity the cognitive authority of science has declined while the technological power of science has vastly increased (RTS 188, 191). As science is increasingly contextualized, a fundamental paradox is encountered—on the one hand, apparently, an alarming decline in science's ability (and authority) to define the reality of the natural world; on the other hand an unprecedented increase in its power to manipulate and intervene in that world. This paradox is all the more remarkable because scientists, by means of the scientific-technical instruments at their disposal, still have, if no longer a monopoly, at any rate privileged access to ways of understanding and manipulating the natural world. Under these conditions the cognitive authority of science might have been expected to be reinforced. However, the diffusion of a wider range of sophisticated instrumentation into workplace and home, the development of global communication and information infrastructures and networks and the growing emphasis on the potential of user-producer relationships as a key arena for the generation of new ideas and applications on which scientific-technological novelty crucially depends are constantly extending—democratizing and commercializing—access to ways to understand, and therefore manipulate, the natural world. Although we are not all "scientists" yet, many more of us are. (RTS 184–85)

Indeed, this paradox seems to signal that modernity and modernization are no longer traveling the same route. One "decisive break in the triumphant success of science is the de-coupling of science's useful outcomes, for which it is now most highly valued, from its cognitive authority. This is a more complex phenomenon than the division of labour, and dependence, implied by the traditional distinctions between pure and applied science, science and technology or research and development. It has more in common with the uncoupling of modernization from modernity, the concrete processes of innovation and improvement from the values on which these processes were once assumed to rely" (RTS 184). The successes of modernization no longer justify modernity's philosophy of science. Instead, the cognitive core of science seems to be emptied of any single authoritative set of ontological or epistemological principles. No longer can science promise certainty, or something close to it, about features of the natural world. But its technologies can indeed promise the "power to manipulate and intervene in that world" (RTS 184–85).[7]

What has caused this sorry state of science's cognitive authority? Its decline has been created by the rise and persistence of epistemological dissonance and dynamism—by controversies. Here RTS points to Steve Fuller's observations in his book *Science* about attempts to standardize scientific knowledge across different cultures. The producers of this knowledge want to standardize it. Yet the consumers "favor forms of knowledge that capitalize as much as possible on what they know already and what they want to achieve" (RTS 186). Thus cross-cultural transfer of knowledge strips it of "its metaphysical and culturally specific elements," reducing it "to its utilitarian essence" (RTS 186). "Seen in this light, scientists' assertion of their cognitive authority is nothing more than a 'market' device to increase the price that knowledge consumers must pay, in both the political and commercial arenas, and also to restrict conditions under which these knowledge transactions take place" (RTS 186–187). Here, too, one can see the two ways to interpret what has happened to the epistemological core of science; either it has been emptied through cultural transfer, or it is crowded with heterogeneous epistemologies (RTS 187).

De-monopolization

As a consequence, scientists, who became the "secular priests of modernity" (RTS 195), or perhaps even the "shock troops of modernity" (RTS 190), have now lost their monopoly on the production of scientific knowledge. Scientific expertise has been distributed to all kinds of groups of people as scientific knowledge continues to be further tested in the "agora" of public contestation after it leaves the lab. The "contextualization" of science is a demonopolization of its control by only scientists. This theme is familiar from Beck's arguments, and as a goal of Latour's.

In summary, GNS have provided a bold argument about the changing nature of scientific work processes and their consequences for science, philosophy of science, and modernity. It is one unlikely to warm the hearts of Mode-1 scientists and their conventional institutions. It seems both descriptive of changes that these authors propose have already taken place, yet also prescriptive in several ways. It approves of the increased reliability of results of research which was promoted by scientists' abandonment of the old "segregated" ideal of research. It sees the increased contextualization of scientific research as already democratizing science in the sense that it must now enter the "agora" and that it thus distributes expertise, thereby offering the possibility for democratizing effects on Mode-2 society as well as Mode-2 science. Yet these are only possibilities, since at some points GNS seem pessimistic about the effects on social equality of Mode-2 science and society.

NPK provided a gloomy prediction that this new way of producing scientific knowledge would in fact increase social inequality. This assessment has been supported by other studies of the phenomenon on which they focus, and even by ones with different analyses of the nature and causes of these shifts (Kleinman and Vallas, "Science, Capitalism"; Klein and Kleinman, "Social Construction of Technology"). In RTS they are more ambiguous about the future. On the one hand, the co-evolution of science and society, apparently undirected by any intentional human activities, continues to seem to promise increased inequalities. Yet they go to some effort to articulate the need to block such a future, for example, by appealing to the assumption of responsibility by scientists for thinking and consulting more about the consequences of scientific projects and by science policy analysts for "including those likely to be implicated" (RTS 255). Moreover, they refuse closure to their own account.

6. DEBATING POINTS FOR THE AGORA

In keeping with GNS's commitment to the importance of lack of closure for particular scientific projects—the positive value to be gained from expecting continual reinterpretation and renegotiation as new groups and new circumstances raise questions about existing de facto "closures"—the concluding chapter of RTS proposes seventeen contentious and unresolved issues requiring further discussion and reflection. Readers will have met these all in the earlier chapters. Here they are selected as issues especially in need of continuing public discussion. I summarize just a few to give a sense of how useful these (and their accompanying short paragraphs of explanation) would be in teaching contexts in the sciences and in philosophy and science studies to stimulate fresh thinking.

Not only the applications but also the unknowable implications of scientific research must be incorporated in that research from its start; we need strategies for accomplishing this (#4, 253). As a start, what are "strategies for exploring more accurately the implications of knowledge production" (#5, 254)? Scientific research is always done in rich and varied social contexts. How can attention to these contexts "be internalized by researchers as a core responsibility" (#6, 255)? Since "there is no scientific or technical problem that is 'only' technical or scientific," attention to the contextualization of scientific research requires subjective experience to be taken more seriously (#10, 257). "The more open and comprehensive the scientific community the more socially robust will be the knowledge it produces. This is contrary to the traditional assumption that there is a strong relationship between the coherence (and, therefore, boundedness?) of a scientific community and the reliability of the knowledge it produces" (#12, 258). We need to better link the "images of science" to the actual practices of science since such images, "even populist representations, support rather than detract from the body of knowledge produced by research." This claim challenges the fear that "lay popularization of science may encourage unwelcome 'lay' interventions (unless carefully managed as part of a campaign to increase the public understanding of science)" (#14, 259–60). Finally, they close the book with the argument that the contentiousness of lay participation in the production of scientific knowledge needs to be rethought. On the one hand, it "is not to be taken as a free entry ticket into an inchoate and unstructured arena of endless (and often futile) debates." On the other hand, "the often feared 'contamination' of science through the social world should be turned around. Science can and will become enriched by taking in the social knowledge it needs in order to continue its stupendous efficiency in enlarging our understanding of the world and of changing it. This time, the world is no longer mainly defined in terms of its 'natural' reality, but includes the social realities that shape and are being shaped by science" (#17, 262).

These bold and ambitious analyses in NPK and RTS stimulate reflection on conventional ways of conceptualizing actual and ideal relations between science and society, as well as on the similarities and differences between their ways of thinking about such issues and how these issues appear in Latour's and Beck's writings. Clearly their descriptions and prescriptions will be deeply disturbing for different reasons to different groups.

7. PROBLEMS

Did Mode-1 science ever exist?

The New Production of Knowledge was criticized for taking the rhetoric of science as the Mode-1 reality (for example, Godin, "Writing Performative History"; Fuller, "Review").[8] It was clear that the authors had overemphasized the break between Mode-1 and Mode-2 knowledge production since the features on which they focus in Mode 2 can also mostly be found in Mode 1. Perhaps these features have become more prominent recently, but multidisciplinary teams, mission-directed science (not just seeking technological applications), transdisciplinary models, and a significant degree of integration in social surroundings are all familiar phenomena to historians of the first three centuries of modern Western scientific research. Think, for example, of the long and ongoing histories of military-funded scientific research and of medical, health, geological, and agricultural research, which have always exhibited such features. Thus NPK took the rhetoric of Mode 1 of modern science for its reality. This problem is acknowledged early in RTS: "The implication of our argument [in NPK] was that science could no longer be regarded as an autonomous space clearly demarcated from the 'others' of society, culture and (more arguably) economy. Instead all these domains had become so 'internally' heterogeneous and 'externally' interdependent, even transgressive, that they had ceased to be distinctive and distinguishable (the quotation marks are needed because 'internal' and 'external' are perhaps no longer valid categories). This was hardly a bold claim. Many other writers have argued that heterogeneity and interdependence have always been characteristic of science, certainly in terms of its social constitution, and that even its epistemological and methodological autonomy had always been precariously, and contingently, maintained and had never gone unchallenged" (1–2). Yet I would argue that the overemphasis on the differences between Modes 1 and 2 persists in the later account. They emphasize how in the past modern sciences conducted a monologue "at" society. In both books they miss how those scientific projects and their monologues were in fact always already responses to how their social environments were speaking to them. In particular, the "autonomy of science" rhetoric, which reached its heyday in the United States along with modernization theory and Mode-1 science in the period immediately after World War II, clearly was an attempt by the leaders of the scientific establishment to protect science from the kind of public scrutiny and accountability which such huge federal expenditures as the newly founded National Science Foundation and the vastly increased investment in scientific research and development signaled by the Manhattan Project would seem to demand. Science didn't want Congress "interfering" in how it spent the funds Congress provided (Hollinger, *Science, Jews, and Secular Culture;* Godin, "Writing Performative History"). This kind of claim to Mode-1 science clearly was a rhetoric with a political purpose, not a description of the reality of scientific research.

However, even if they overemphasize the break between Modes 1 and 2—even if Mode 1 never existed in any form but a rhetoric—I think their discussion of Mode 2 is valuable. They draw attention to aspects of the contemporary situation of science and modernity which are not fully grasped, or not set in the valuable contexts articulated here, in either the other science studies accounts discussed earlier or elsewhere —that is, apart from the echoes of their arguments in Beck and Latour. There have been important changes in the trajectories of both science and society in the last few decades, and GNS identify some significant aspects of those changes.

Society or politics? Power relations in modernity, modernization, science, and technology

In both books there is a disturbing tension between, on the one hand, their overt concern to identify and envision a democratic context for the production of scientific knowledge, which could decrease the social inequality which they see as even more a consequence of Mode-2 scientific practices, and, on the other hand, the absence of any discussion of the political and economic obstacles to such a project apart from their frequent observation that scientists tend to resist such contexts and processes. There is no discussion of how power relations of race, class, gender, and imperialism have already shaped the sciences and technologies we have and the societies that have co-evolved with these sciences and technologies. In their discussions of the contextualization of science and of "social accountability" the interests of corporate capital and national security, on the one hand, and social justice movements, on the other hand, are lumped together as "contexts" for the production of scientific knowledge and technological power and as social "voices" that "speak back" to science. While these are all indeed contexts for the production of scientific knowledge, crucial differences between them are lost with respect to both the kinds of sciences such different interests want and the kinds they have the power to get. We do not hear how disempowered groups get to be heard when they lack the economic and political power to capture governmental or media agendas. We do not get to hear how a democratic "agora" can come into existence. After all, an agora is not inherently structured in any particular political way. Recollect that classical Athens, from which the term "agora" comes, was a slave-holding and male-supremacist society with only a tiny proportion of its population in the category of citizens.

Their accounts annoyingly use the depoliticized neo-Liberal language of management and policy—of corporate capital—to discuss situations of radical and increasing inequality. In both books they adopt a relatively benign stance toward the new modes of knowledge production. Of course they do acknowledge in both books that Mode-2 production of scientific knowledge seems to promise increased inequality, mostly, they say, because it narrows access to the results of research. Others have argued that the widespread praise for the increase in new knowledge workers, such as one finds here, obscures its shadow of increases in exploited labor at the very sites where the new knowledge is being produced (cf. Kleinman and Vallas, "Science, Capitalism"; Sassen, *Globalization and Its Discontents;* I return to this issue in chapter 8). One could reasonably see GNS as promoting the increased power of corporations and states over the production of knowledge as universities and their traditional ethical and relatively democratic social values decline in power. One could see them as only trying to improve the public image of sciences which are increasingly under corporate and state military control (cf. Godin, "Writing Performative History"; Linder and Spear, "Review of *Rethinking Knowledge*").

Thus the focus on "society" rather than the "politics" centered in Latour's and Beck's accounts turns out to be significant.[9] One would have to conclude that they are not interested in transforming the realm of the political—or, at least, not in thinking about it in the policy contexts of their studies. But should we want, let alone trust, science policy which cannot engage with issues of the political inequality which, according to GNS, it both generates and subsequently supports?

Two final points. Both the language of evolution and the management framework which they use work to establish the kind of Mode-1 authority for their own account which their argument is positioned against in the case of their subject matter: "science." The evolution language naturalizes and depoliticizes linked changes in science and society past, present, and future. There are no intentional social agents for these "evolutions." What would be the point of political organization against or in support of a natural process? The management framework, too, appeals to technical, expert, top-down solutions to social challenges. Of course in other ways they support the activities of the political domain of the agora in shaping science and modernity/modernization. So their attachment to these discourses is puzzling.

Androcentrism and Eurocentrism: Other contexts, others speaking back

Two kinds of politics have been centered in this book as important to our attempt to understand science and modernity. These are feminist and postcolonial politics. According to GNS, apparently neither these nor their critical targets—male supremacy and Eurocentrism/colonialism/imperialism—should be considered significant shapers of sciences and modernities, past, present, or future. The story they tell is virtually entirely about European, bourgeois men, who for the most part act as individuals, or collectives of them, though they characterize their projects as about "humans." (Recollect their admonition that we must pay attention to how "people" enter science [RTS 246].) Almost no women, non-Europeans, or poor people appear as either agents of history or agents of knowledge, nor do they appear as the unfortunate recipients of most of the costs of Mode-1 and, probably, Mode-2 sciences and societies.

This is not to say that the authors are totally unaware that humans come in such "marked" categories. In the case of women, they do report feminism drawing attention to the absence of women in science, and as objects of study, for example in clinical trials. They attribute to university women's studies programs one of the most successful examples of university-community cooperation in scientific projects (RTS 140). They point to the value of the women's health movement in "initiating research on otherwise neglected problems" (RTS 213). They praise feminists for their criticisms of objectivity and neutrality of science (RTS 212). And they attribute to feminism the articulation of the need "for accountability of science to a wider public that need[s] reassurance on the proper working of its quality control system" (RTS 60). That is, the idea which is central to their own project that the more socially robust a science is, the more reliable are its claims. This is definitely an improvement over the absence of attention to feminist work in Beck and the implicit delegitimation of feminism as an inevitably regressive "identity politics" in Latour.

Yet this attention to feminist accounts and the women's movement are brief and scattered, occupying perhaps a total of three pages in the RTS book of 262 pages. Moreover, they persistently seem confused in these accounts. For example, what are we to make of the following passage? "Of course, people are always present in our knowledge, be it as the objects of study and research or as those who are active in doing the research. But it was largely due to the insistence of the feminist movement that this distinction gained in relevance. It is now much more widely accepted that the human objects of research—be they gendered or in other ways socially categorized—have to be carefully attended to, if research results are not to be unnecessarily restricted in their validity or wilfully distorted in their applicability. In clinical research especially, topics like 'gender differences in metabolism' (or in other biological functions) have opened up new and promising avenues for research" (116). What are we to make of this passage, which conceptualizes gender differences as biological functions and, evidently, of interest only when "socially categorized," that is, when women are the objects of study or researchers? What are we to make of "be they gendered or in other ways socially categorized" when every human is always "gendered" and always in other ways socially categorized—for example, as European, bourgeois, and masculine? Whatever they have in mind here, we are entitled to observe that when women emerge into view, they seem to mark a site where the authors no longer can think clearly. Suddenly the world becomes unintelligible! (I return to this phenomenon in chapter 8.)

Moreover, though they continually identify problematic contrasts of modernity and its philosophies of science, they never identify such contrasts as themselves gendered, classed, or imperialized (orientalized); the "context of implication" is not invoked here. Thus we hear about the overvaluing of the "hard" scientific and epistemological core of Western sciences and their modernity vs. the "soft" social and cultural worlds with which such cores do in fact continually interact. They never speak of the specifically masculine, bourgeois, and Eurocentric character of the sciences, past, present, or, evidently, future. They never see "the modern" or "the scientific" as gender, class, and imperial horizons intended to prevent thought from straying past these boundaries into the dangerous incomprehensible terrains of such social and political histories and present practices.

Yet modernity does function as just such a horizon in their own work. One way this occurs is in their nomenclature. Mode 1 begins with the philosophy of the seventeenth century and the practices of industrialized science. It ignores, or rather, excludes the craft production of scientific knowledge, in contrast to the kind of production processes often referred to as "industrial," that is, characterized by a hierarchically organized division of labor with control of the work process retained by its managers. The industrial production of science emerged only in the late nineteenth century and the twentieth, and in its university disciplinary location characterizes much of what NPK refers to as Mode 1. Craft production still does and must persist at early and, some would argue, ongoing stages of the formulation of any kind of scientific project. Consider, for example, Watson's, Crick's, and Franklin's work recorded in Watson's *The Double Helix* and in Sayre's *Rosalyn Franklin and DNA*. It certainly characterizes the production of scientific knowledge in many other cultures around the world. I would prefer a "craft," "industrial," and "postindustrial" set of categories to avoid the assumption that science begins only when European sciences and later North American sciences are the topic and they reach an industrial stage of production, and that science is co-extensive with only this production model. Their nomenclature reinforces the exceptionalist and triumphalist Eurocentric and androcentric view that modernity and its sciences are by definition Western modernities and sciences.

Furthermore, though they assert the importance of processes of decision making and social/scientific transformation which are open to input by all of those affected by them, they do not in fact avail themselves of the content of feminist and postcolonial science and technology studies or the broader fields of feminist studies and postcolonial studies within which the science and technology studies concerns are articulated. Women and "postcolonials" remain silent and, indeed, virtually invisible in their accounts, though their voices and ideas have been widely available for the last couple of decades. These groups have some useful strategies to achieve the kinds of goals GNS espouse, and they have additional goals not considered by GNS. Of course Latour and Beck also do not bring this work to bear on their analyses. These science studies scholars all have good intentions. Yet the global needs and desires represented in feminist and postcolonial science and technology studies need more than good intentions from sciences, modernities, and science studies.

8. CONCLUSION

In spite of its limitations, this ambitious project does offer valuable resources to those interested in the democratic transformation of sciences and their societies in the West and elsewhere. Perhaps most notably for our project here, it delineates in compelling general terms just how it is that the more socially robust one's knowledge claims, the more empirically reliable they will be. That is, the more scientific research projects engage with their social environments in egalitarian discussions, the higher the quality of the results of that research. Such an analysis can be interpreted as generalizing precisely the kinds of standards for good research that feminist and postcolonial science studies have provided. These authors also ingeniously translate the still valuable parts of the local vs. universal sciences contrast into the more relevant contrast between socially integrated vs. socially segregated sciences. GNS have found a way to make generally compelling the importance of a reversal of the desirability of these phenomena from the way conventional philosophies of science conceptualized them. These two issues have been especially difficult ones for conventionalists to grasp. Resistance to such reversals have fueled the Science Wars and attacks on science studies, and especially on feminist and postcolonial philosophies of science. Such resistance has bolstered the empirically unsupportable view that it is reasonable to find feminist and postcolonial science and technology studies irrelevant to the production of good science.

We turn now to standpoints on science and on modernity from peoples living on the horizons of the Enlightenment. What can we learn if we start off thinking about science and modernity from those lives?

II.

VIEWS FROM (WESTERN) MODERNITY'S PERIPHERIES

4.

WOMEN AS SUBJECTS OF

HISTORY AND KNOWLEDGE

— — — —

THIS PART OF THE BOOK will not follow the model of the first part, in which each chapter discussed an influential representative of a particular strain or legacy within postpositivist science studies. Here we will look at contributions from many participants in each of three kinds of social movements representing groups at the periphery of modernity: Western women's, postcolonial, and Third World women's science and technology movements. There are several reasons for this plan.[1] For one thing, influential as some of the individual researchers, scholars, and activists in these movements have been, virtually none holds the kind of powerful institutional position—in universities, science policy circles, or professional organizations—occupied by Latour, Beck, and members of the Gibbons, Nowotny, and Scott team. It is these movements which have been most influential, in my opinion. In the second place, the emphasis in all three of these movements has been on the intellectual and political importance of multiple voices. Focusing on only one representative of each group would do a disservice to the many scholars whose work would be marginalized by such a practice. Moreover, at least in the feminist movements, the ones selected could also end up unhappy at having to face criticism about their "star" status which such an account would de facto promote. Finally, many of the insights identified in the accounts which follow have, in fact, emerged more or less simultaneously from multiple sites within social movements. Consequently they are collective achievements as well as individual contributions to a degree far greater than those in the analyses of Part I.[2] I don't mean to intimate that there are no individual insights in these three fields which have rightly achieved a distinctive authority. To the contrary, quite a few have done so within her particular field—though not, with some exceptions, in the other fields. Rather, I wish to draw attention to the different conditions for the production of scientific knowledge which have tended to be characteristic of work emerging from these increasingly powerful social movements.

We can begin with feminist science studies (fss), which emerged in the 1970s at the conjunction of the activist Northern women's movements and the new field of postpositivist science studies.[3] How do feminist challenges to the standards of "good science" provide resources for the interrogation of the modernity/tradition binary itself? To what extent is the category of "the modern" problematized in this work? What is the relation of this work to the exceptionalism and triumphalism which have plagued even the work of those mainstream science studies scholars who overtly reject such stances?

Scientists, activists, and scholars working on health and environmental issues were the most energetic groups initially raising such issues. However, physicists, philosophers, historians, ethnographers, and sociologists of science also articulated their concerns in this new field from the beginning. Engineering, chemistry, and mathematics have subsequently been analyzed.[4] This work has had the audacity to go beyond issues of inclusivity to challenge the technical content and standards of empirical and theoretical adequacy of Northern sciences. It has remained both highly acclaimed and also controversial, including among feminists themselves. A lot more than "a fair chance for the girls" turns out to be at stake in these kinds of criticisms of the jewel of scientific rationality which sits in the crown of modernity and of modernization theories.

To answer the questions posed about fss we will have to consider issues not only about the sciences, but also about the politics within which sciences are constituted, which, in turn, help to constitute political philosophies and projects. Section 1 of this chapter provides a brief overview of fss concerns of the last three decades and begins to assess the extent to which they contribute to rethinking both the sciences and politics of (Northern) modernity. Section 2 pauses to identify how the term "gender" in feminist work means much more than simply the identity of individuals as boys or girls, women or men. Its broader use enables distinctive insights about modernity and its sciences and politics. Section 3 looks at how feminist standpoint approaches provide a logic and methodology for sciences "from below." Section 4 identifies the strengths of FSS in general and standpoint theory in particular in challenging conventional understandings of modernity. The concluding section 5 identifies some limitations of this work.

One of the limitations should be mentioned immediately: the Northern studies, with important exceptions, do not completely abandon the exceptionalism and triumphalism of conventional philosophies of Northern sciences and modernities insofar as they do not take the standpoint of women in non-European societies on either Northern or indigenous sciences. (Of course, the Northern sciences, philosophies of science, and history and social studies of science, from which the feminist science studies scholars in part originate and to which they today professionally belong, also are deeply Eurocentric in such respects.)[5] Southern feminist science studies will be discussed in chapter 6, after we have identified distinctive concerns of the field of postcolonial science and technology studies.

1. FIVE CONCERNS OF NORTHERN FEMINIST SCIENCE AND TECHNOLOGY STUDIES

Discrimination against women in the social structure of the sciences
Complaints about discrimination against women in scientific, educational, professional, and state institutions date back at least to the mid-nineteenth century in the United States and Europe (Rossiter, *Women Scientists in America*; Schiebinger, *The Mind Has No Sex?*). By the 1970s in the United States it became illegal to discriminate against girls and women in education, publishing, jobs, scholarships, membership in scientific societies, and all the other venues which had vigorously defended their ramparts against the presence of women as scientists.[6] New histories of women's struggles for inclusion in scientific institutions began to appear. Outreach and affirmative action programs were developed, and some progress was made.

Yet a generation later, such concerns continue to remain relevant, unfortunately. Public figures such as the president of Harvard University still have found it appropriate, in 2005, to wonder in public, in spite of long accepted evidence to the contrary, if the low number of women in science at institutions such as Harvard is best explained by biological differences between the sexes.[7] Meanwhile, even the highly distinguished young women scientists at MIT, across the street from Harvard, report difficulties integrating domestic with professional responsibilities—a challenge their colleagues do not experience. Their senior women colleagues reflect on the fact that though they cannot say that they have experienced discrimination (after all, they are at MIT), processes mysterious to them seem to leave them with fewer departmental resources than their male colleagues.

As long as discrimination against women is prevalent in the larger society, an end to legal discrimination will not be sufficient to enable women to get access to the full array of resources and opportunities which are available to male scientists. Or, to put the issue another way, patterns of institutional practice and of scientific culture which are considered ideal from the vantage point of the lives of men in Western professional classes who have designed such practices and cultures will inevitably discriminate against people with other kinds of social responsibilities and desires. Not everyone thinks that becoming a Western professional man represents the height of human achievement. Of course "Western" science now has deep roots in many other cultures around the globe, and patterns of professional life in even a field like high energy physics are not identical in, for example, the United States and Japan (Traweek, *Beamtimes*). Innovative recent studies are exploring the deep and subtle ways in which becoming a scientist and becoming a man tend to be linked, and the resistances to such processes in which girls who love science and women scientists engage (Brickhouse, "Embodying Science").

To get a larger perspective on this issue, the United States is by no means in the lead in placing women high in the governance of scientific policy or academic research. For example, United Nations data show that eighteen nations had a higher percent of women in physics departments in 1990 than did the United States. Most of these are Eastern European and Third World states (Harding and McGregor, "Gender Dimension" 12). Were those countries more feminist than the United States? No, and the reasons for the greater participation of women in physics were different in different countries.[8] At the same time, in many parts of the world there have been widespread efforts to increase scientific literacy among girls. The conditions which made life at MIT difficult for junior women scientists also make any kind of schooling difficult to achieve for the vast majority of girls and young women in the Third World who have responsibility for child care and domestic work. In such circumstances, girls tend to end up education-ally deprived. This is especially a challenge in poor sectors of societies, where many workers are required to generate sufficient resources to enable families to survive. Thus large families are economically valu-able, and young girls' labor in child care and domestic work is neces-sary. (We now know, of course, that it is poverty that causes overpopula-tion, not the reverse, as international population policy assumed for so many decades.) If a family can scrape together enough money for one child to go to school, boys' education will be favored over that of girls. Yet the education of women is the single factor most responsible for reducing birth rates, according to demographic studies. Increasing the education of girls is a tangled but not unsolvable challenge.

Thus one should not be surprised to find the absence of women in science and engineering educations and careers to be intertwined with broader social policies and practices. What looks like a relatively con-servative feminist demand for equal education and employment turns out to require widespread and deep transformations of social relations. These demands challenge the exclusion of women from the public sphere. Insofar as this issue focuses on women's active agency in the design and practice of science, it ends up being about far more than merely adding women to scientific work sites. It interrupts the exclu-sive association of rationality, objectivity, and social progress with only the activities of men in the dominant groups. It opens the door to reevaluating for whom—for which social groups—standards of ra-tionality, objectivity and social progress have been formulated. In cases where the inclusion project also articulates women's needs from the public sphere, such as the provision of child care and an equally valued career path that allows for family responsibilities, it presents an even more radical challenge to the ways public institutions, including scien-tific and technological ones, are ideally envisioned. And the presence of increasing numbers of women in science institutions in turn prom-ises powerful resources for anti-sexist social change more generally. Of course, it represents an important inroad in the modern separation of gendered public and private spheres. In doing so, it also lays ground-work for the emergence of women as *agents* of knowledge and history, as new kinds of historical and scientific speakers or *subjects*—that is, as the kinds of subjects modernity has forbidden, as we will see in chapter 8.

Sexist sciences

A second focus of feminist concern has been to identify and criticize the production of sexist results of scientific research in even the most prestigious and influential work in biology and the social sciences. This "sexist science" has persistently attempted to document and ex-plain women's social inferiority in ways that justify male-supremacist discrimination against women as natural. The sciences have thereby provided powerful resources for the continued oppression of women and exploitation of their labor, including their caring labor. This kind of natural and social science has been criticized since the late nineteenth century (cf. Fausto-Sterling, *Myths of Gender*; Schiebinger, *The Mind Has No Sex?*; *Nature's Body*). It has diminished in recent decades as institutional recognition of its faulty evidence and reasoning in biology and the social sciences has vastly increased. But it certainly has not yet completely disappeared; we are subjected to attempts to reinvent it in generation after generation. Why does this happen?

Both the feminist biological and social science work on this issue has been widely disseminated throughout the media and makes constant appearances in popular public discussions (though usually not with the word "feminist" attached to it). Indeed, there could be no better evi-dence that the "truth" alone does not set people free than the fact that more than a century of such influential scientific counters to male-supremacist assumptions has not succeeded in eliminating either ac-tual discrimination against women or the still-far-too-widespread as-sumption that there is nothing wrong with the president of one of the premier U.S. universities today intimating in public that the low per-centage of women in highly prestigious positions there could possibly be caused by their biological inferiority. Evidently more is at issue here than the false beliefs on which Liberal political philosophies have focused. Psychological fears of women's "powers" and economic/political interests in their exploitation are two candidates for that "something else." Neither of these has been a central focus in the largely Liberal mainstream of U.S. and European feminist science studies. The first calls up terrifying phantasms of Freudianism and the second of Marxism, both of which have been anathema to mainstream postpositivist U.S. (at least) philosophy of science and not obviously welcome in the rest of postpositivist U.S. social studies of science (Harding, "Two Influential Theories"). For better or worse, I am far less terrified of these two intellectual giants and the resources their legacies can provide for feminist and postcolonial understandings of modernities and sciences.

Applications and technologies

Feminist science and technology projects also have drawn attention to the sexist and androcentric applications of scientific theory and information, and the sexist and androcentric design and maintenance of scientific technologies. Reproductive theories and technologies have been a continuing focus, but attention has also been directed to domestic and workplace technologies, architecture and urban design, and the whole larger field of technology design, repair and maintenance (Wajcman, *Feminism Confronts Technology*; *Technofeminism*; Oldenziel, *Making Technology Masculine*).

Two advances in mainstream postpositivist work permitted valuable expansions of feminist issues here. One was the emergence of social constructivist understandings of technology (MacKenzie and Wajcman, *Social Shaping of Technology*). The older positivist accounts had insisted that the term "technology" refer only to material artifacts or hardware, which could be presumed to be value-neutral and culture-free. Yet the constructivist approaches revealed the misleading character of such conceptions by focusing on how technological change became a site for social struggles over not only the nature of the artifacts, but also how the benefits and costs of the new technologies would be distributed; who would have the knowledge of how to design, use, and repair the artifacts; what would be the cultural meanings of the artifacts, and the resources and costs that such meanings produced. It turned out that gender struggles often occurred at sites of technological change.

The other advance was the parallel way the social constructivist tendencies in science studies showed how economic, political, social, and cultural elements permeated even the cognitive, technical core of the sciences themselves. Where the positivist theories had demanded a demarcation between "pure sciences," or "basic science," and the applications which the results of such research made possible, the new studies showed how there were no defensible boundaries between the two processes. Even "basic science" can be mission-driven; consider, for example, how this is so for medical and health research or environmental research.[9] Thus feminists could argue that gender elements, too, appeared in the missions to which even basic science responded. Why else were women's health challenges researched less than men's in the cases of heart attacks, neurological challenges, and cancers affecting both men and women, as studies by the United States National Institutes of Health highlighted? Why were diseases such as AIDS not defined in ways that permitted their detection in women, too? Recently, accounts of the deeply intertwined relations between sciences and societies have emerged in every subfield of science studies. So feminist concerns about the social character of science reinforced such concerns about technologies and vice versa.

Science education: pedagogy and curricula

A fourth focus of FSS has been on science education. It turns out that making science educations formally available to girls and women is only a first step. Researchers have identified how it is not purportedly deficient girls and women who are responsible for women's low numbers in courses in many scientific and technological fields. Rather to blame are deficient pedagogies and curricula through which girls and boys are recruited to and then engaged in science education. Girls and women don't care about or even are resistant to finding desirable many of the obvious uses of modern sciences and technologies, for example, for profiteering medical and health empires and for militarism. They are less tolerant of dissecting frogs, depriving poor children of the health care they need, environmental destruction, and building missiles than are their brothers, though of course many boys and men object strongly to such practices. Moreover, teachers' expectations that girls will conform to middle-class standards of femininity tend to discourage from achievement in the sciences many working-class girls and girls of non-white races and ethnicities who in fact start out in elementary school loving science and technology. Becoming a scientist turns out to be suspiciously linked with becoming a certain class- and race-preferred sort of boy or man, but not with the kinds of girls or women which the dominant cultures project as desirable. Classroom-preferred forms of femininity are dangerous to girls' educations! (Brickhouse, "Bringing in the Outsiders" and "Embodying Science").

Philosophies and epistemologies of science

Finally, feminists have focused on the distorting gender lenses of standard epistemologies, methodologies, and philosophies of science, visible in scientific practices as well as in histories and sociologies of science. Such issues about the standards for "good science" appeared in feminist science studies from its beginnings. Here the argument has been that gender relations have shaped not just who gets to do science, but also the content and philosophical framework of even the most highly regarded sciences. That is, it is not just "bad science" which bears such androcentric fingerprints, but worse, even "good science." The consequences of this androcentrism are bad for social justice. But they also deteriorate the adequacy and, thus, legitimacy of scientific claims themselves. The very best of Northern scientific work, these critics claim, appears to lack the resources necessary to identify and engage with some of its deepest and evidently most passionately experienced social commitments, such as to male supremacy. This science is epistemologically underdeveloped insofar as it cannot detect how androcentric commitments can, and all too often do, shape every stage of the research process, from the selection of "interesting" problems and useful concepts and hypotheses, to the design of research, the collection and interpretation of data, and the standards for evidence and for what counts as reasonable and convincing results of research. Feminist science studies has proposed scientifically more competent and politically more progressive standards for objectivity, rationality, good method, and "real science." These analyses have been on the frontier in science studies in the North, often virtually alone in demonstrating that epistemologies, methodologies, and philosophies of science can have at least as devastating discriminatory effects as more obvious sexist, racist, or class practices (cf. Code, *What Can She Know?*; Haraway, *Primate Visions;* Harding, *The Science Question in Feminism* and *Thinking From Women's Lives*; Harding and Hintikka, *Discovering Reality*; Keller, *Reflections on Gender and Science*; Longino and Keller, *Feminist Philosophy of Science*).

The field of feminist science studies is far richer and more illuminating than such a brief overview can convey. It will be worthwhile to take time to get a more detailed understanding of one of its concerns which is especially relevant to our project here. One of the most influential feminist epistemological and methodological projects has been feminist standpoint theory. It remains highly controversial, even among feminists, though its use as a methodology is now widespread in biology and the social sciences (Harding, *Feminist Standpoint Theory Reader*). Its insights and practices are crucial for anyone concerned with transforming the intertwined theories and practices of science and society. However, before identifying the distinctive arguments of standpoint theory in section 3, let us first review what is (or should be) meant by the term "gender," which appears throughout all feminist work.

2. WHAT IS GENDER?

I cannot assume that readers know what this now-familiar term refers to in feminist research and scholarship. The term frequently is used in popular writings in ways which conflict with the uses to which feminist studies have put it. Moreover, in a number of languages, in Europe and elsewhere, it has no exact translation, and so the English word has been adopted or some other kind of accommodation to the ideas it expresses has been made. Three popular usages are particularly problematic. It is used to refer to biological differences (sex differences); to women alone; or to some purportedly timeless, universal properties of women. Let us identify some distinctive characteristics of "gender relations" in the feminist work.[10]

First, gender relations are conceptualized as completely social relations. "Women are made, not born," as Simone de Beauvoir famously put the point. Masculinity and femininity are not attached to sex differences in fixed, discrete, or universal ways, though gender assumptions do shape reproduction patterns, and they shape what counts as purportedly normal sex differences. (Of course what is normal is itself a matter of controversy.)[11] Anthropologists and historians report that for almost every attribute or behavior regarded as masculine or feminine in one culture, one can identify a culture in which it is marked the other way. So there is no particular configuration of gender characteristics that marks a woman or a man in every culture. Woman and man, femininity and masculinity, do not have essences.

Moreover, to state the obvious, gender is not just about women, but also about men and, most importantly, the social relations between them. One cannot understand how femininity principles, practices, and values function in any particular cultural context without also examining how masculinity functions there. To talk about "women and modernity" can obscure the need for a discussion of men, masculinity, and their relations to modernity. Thus gender does not become a relevant social variable only when women walk into a room. Men's relations with each other, in board rooms, tribal councils, work sites such as farming, laboratories, professional meetings, governmental agencies, militaries, priesthoods, as well as in the household and domestic life, are also gender relations. Consequently projects to change gender relations must focus on changing men too, not just changing women (see Connell, "Change among the Gatekeepers").

The tendency for Westerners, like other people around the globe, to assume that gender relations are natural and therefore universal has resulted in great harm to other cultures. This occurs as Western policymakers, administrators, or local workers in development or humanitarian agencies, usually intentionally, import Western gender stereotypes and androcentric practices into these cultures under the false impression that these are "natural" or, at least, the only desirable gender relations. Of course one does not have to travel far to find this kind of policy and practice, as we saw in the studies of how white, middle-class norms of femininity in U.S. classrooms excluded girls of color from promising careers in science. And feminists themselves have often made such essentializing generalizations, even if they did not have natural causes in mind. This has prevented them from grasping distinctive gender issues in different cultures (Narayan, "Project of a Feminist Epistemology"; *Dislocating Cultures*; Mohanty, *Feminism Without Borders*). This insistence on gender relations as a cultural phenomenon does not deny that there are biological differences between males and females—obviously there are! Rather, it argues that the "biological facts" are always open to cultural interpretation. More accurately, they are co-constituted with gender relations as sciences and societies change in various ways.

Second, the social relations of gender are hierarchical. Women's and men's work and personal characteristics may or may not be complementary or claimed to be of equal importance; they are always hierarchically organized, though to different degrees in different societies. According to anthropologists and historians, it is hard to identify any universally gendered social characteristics. However, even in the most egalitarian of societies, women are less likely to be found as the leaders or rulers of groups composed exclusively of men, and the more powerful the group, the less likely it is to have a woman leader.[12]

Third, it is important to see that in feminist accounts gender is being treated as a property of three different kinds of social entities. It is a property of individuals; we all, or at least most of us, are assigned a gender, girl or boy, woman or man. Yet it is a property of social structures; some institutions and societies have "more gender" than others according to how widely and rigidly they institutionalize divisions of labor by gender. The great majority of occupations in the United States are gender-segregated in that the majority of such workers belong to one or the other gender. Nursing and university professorships are both considered gender-segregated even though a minority of workers in each kind of job belong to the other gender. Thus militaries, corporations, and public institutions have usually insisted that their managers and administrators be men, and that lots of the low-paid work of maintaining the material worlds of administrators be done by women as secretaries, data managers, office cleaners, food preparers, and the like. Social structures can also be gendered in terms of whose interests they serve: is it men's or women's issues which for the most part shape the ways in which nursing and university professorates are organized and the issues they prioritize? Finally, gender is also a property of symbolic structures. Objectivity, rationality, rigorous observation, and moral insight have usually been gender-coded as masculine. These are kinds of "implications" of the sort that can get attached to scientific research that Gibbons, Nowotny, and Scott discussed in the last chapter (though gender was not specifically discussed there).

Each of these three kinds of gender differences is important. Recognizing them enables one to see the larger picture of how modernity is gendered by looking at the complexly conflicted and coordinated relations of modernity to masculinity and to femininity in particular social contexts. We will look further at the gender coding of modernity and modernization in chapter 8. The point here is that it is especially important to understand that "gender issues" are more than issues about individuals. They are also and more fundamentally about the structural societal relations which assign different groups of individuals to different kinds of lives, and they are about the symbolic meanings of woman and man, masculinity and femininity. Note that one can see these three kinds of gendering functioning separately in some cases where they conflict. For example, it is usually only when women are already starting to make gains in their positions in social structures that a rise in the articulation of misogynous meanings occurs. Social change, especially the fear of loss, tends to produce such discourses.

Fourth, since gender relations are fully social relations, they cannot be historically static, for they must change over time in any given society. They constantly shape and are shaped by other kinds of social relations, such as class and race relations; ethnic, religious, and other cultural relations; and social processes, such as urbanization, industrialization, state formation, or colonialism. Thus a shift in one kind of social struggle will result in changes in other such relations. When suffrage was won for African American men but not women, African Americans became much more interested in women's suffrage. As Western women in the dominant classes see how corporate globalization is designed to exploit women's productive labor around the world, we come to understand better how our expected consumption practices are assigned to match the production practices of transnational corporations (Mies, *Patriarchy and Accumulation*). Some have conceptualized such situations as about how gender is "intersectional" with race, class, ethnicity, and other socially significant structural and symbolic tendencies (Crenshaw, *Cultural Race Theory*). Thus the "unity" of feminisms which some critics call for has come to seem a distinctly undesirable goal for many women around the world. Claims about the desirability and grounds of women's unity seem destined to discriminate against the needs and interests of less powerful groups of women. This fact explains why many feminists think it important that feminisms not be unified. Would we expect to see a global unity of men?! Instead, the possibility of powerful coalitions among women's groups around the world becomes a necessity, not just a luxury, from the perspective of such considerations. Solidarity with other women in their historically distinctive struggles seems a better goal than women's unity.

Note that a commitment to understanding gender as interactive and co-constituting with any and all other social hierarchies, as empirical evidence requires, gives feminist work an internal logic of concern for topics which are central to other pro-democratic social movements. For feminist science studies, attention to how race, class, empire, and colonialism have influenced the conceptual framework and content of sciences around the globe is not just a matter of having a good attitude toward others less politically, economically, and socially advantaged than oneself. Rather it is a requirement for understanding how gender hierarchies themselves function in distinctive historical ways *here*, and for *us*, wherever and whoever "we" may be. This is not to say that feminists in dominant groups always, or even usually, center understandings of gender as interactive with race, class, and other structural features of societies. My point is rather to draw attention to the logic of interaction here, which, when followed, can prevent thinking of "multiple oppressions" as only additive, or as an interesting but not necessary concern.

Finally, while I have been describing gender systems as empirical facts, "out there" in social worlds, gender also is a theoretical, methodological, and analytic lens through which one can examine institutions, their cultures, and their practices, including peoples' culturally specific assumptions and beliefs. Gender is "out there" in the world humans have made in the ways indicated above, but it can also provide a conceptual framework, like class or race, through which to examine phenomena not usually thought of as gendered.

Now we can turn to standpoint theory, which uses such differences as those of gender, race, and class, to provide resources for achieving stronger forms of the objectivity, reliability, and rationality of scientific work than conventional sciences and philosophies of science have produced. Standpoint theory produces the kind of philosophy of science/epistemology and research methodology necessary to provide both empirically and theoretically more competent sciences which can thereby serve to help bring into existence and maintain a radically inclusive democratic social order.

3. SCIENCES FROM BELOW: STANDPOINT THEORY

Origins and purposes

Standpoint theory's development in feminism originated in attempts to explain two things. One was how what was widely recognized as "good science" or "good social science" could produce such sexist and androcentric research results as feminist social scientists and biologists were documenting. The other was to explain the successes of feminist work which violated the norms of good research, such as engaging in research that was guided by feminist politics. In both cases, the theorists sought an epistemology, philosophy of science, and methodology which provided more resources for such projects than did the prevailing feminist corrections to the standard empiricist or positivist approaches. A third reason for developing feminist standpoint theory was methodological and political: to direct attention to how to produce knowledge that was *for* women, not just about them.

Women needed information about their bodies, their environments, and how social institutions worked that the existing natural and social sciences did not think worth pursuing. The early theorists all turned to the work of Marx, Engels, and Lukács and their heirs in the Frankfurt School of critical sociology. The Marxian tradition had identified the importance of the standpoint of the proletariat to social transformation. However, the Marxian legacy seemed unable to overcome various limitations, and had pretty much been abandoned by the 1940s.[13]

Standpoint theory emerged anew in the 1970s and 1980s independently in the work of Nancy Hartsock, "Feminist Standpoint"; Alison Jaggar, "Feminist Politics and Epistemology"; Hilary Rose, "A Feminist Epistemology"; and Dorothy Smith, *Everyday World as Problematic* and *Conceptual Practices of Power.* These scholars came from three different disciplines and three different countries. All four argued that the best of feminist research does and should start off empirical and theoretical research projects "from women's lives." In doing so, this research articulated projects for what could be called "sciences from below." As a specifically feminist project, standpoint theory has since been developed in many areas of the social and natural sciences. Yet it seems also to be an organic epistemology, philosophy of science, social theory, and methodology, emerging in virtually every pro-democratic social movement, whether or not it is specifically named as a standpoint project. It possesses this organic character in the sense that when marginalized groups step on the stage of history, one of the things they tend to say is that "things look different if one starts off thinking about them from our lives."

In its recent explicit articulation, however, standpoint theory has been developed primarily as a feminist project, though of diverse types. Different theorists and groups conceptualize its disciplinarity or "genre" in different ways. Three of the most influential are as a critical theory—a critical political philosophy, social theory, or sociology of knowledge—of the relations between knowledge and power (thus as a kind of epistemology and philosophy of science); as a method or methodology which can guide research projects; and as a political resource which can empower oppressed groups. A number of feminists of color have developed it as a way to describe what can be seen from their distinctive location in gender, race, and class social relations, and as a criticism of research carried out within narrow disciplinary frameworks (Collins, *Black Feminist Thought*; hooks, "Choosing the Margin"; Sandoval, "U.S.

Third World Feminism"). Like the rest of feminist science studies, it shares distinctive features with two larger fields—post-Kuhnian and feminist science studies. Yet it also contrasts with main tendencies in these last two projects in significant ways.[14]

A prescription: Start off research from women's lives

Early feminist social science, health, and biology researchers seemed to deviate from their recommended disciplinary practices in that they did not start off research from the standard theoretical and method-ological conceptual frameworks which their particular disciplines thought important. For example, Carol Gilligan (*In a Different Voice*) started off thinking about the absence in moral theory of attention to the distinctive kinds of decisions women faced as mothers and care-takers. Moral theory elevated to the highest ethical categories only the kinds of decisions which men made as managers, administrators, lawyers, and the like—decision-making from which women had long been excluded. Why was it that the most influential authors on moral-ity and moral development (such as Kant, Freud, Piaget, Rawls, and Kohlberg) could not perceive the kinds of moral decisions with which women were faced as also capable of exemplifying the highest catego-ries of moral thought? Similarly, Catharine MacKinnon ("Feminism, Marxism, Method") identified how what counted as rape and what counted as objectivity had a distressingly close fit with only men's conceptions of such matters—conceptions that reasonably arose from men's kinds of social experiences with women and in institutionalized public thinking, such as in law courts. What happened when women health and biology researchers studied women's bodies, and brought their own experiences of menstruation, orgasm, pregnancy and birth-ing, and menopause to bear on the standard biology and medical ac-counts of such phenomena? (Hubbard, *Politics of Women's Biology*; Hubbard, Henifen, and Fried, *Biological Woman*).

Thus these researchers and scholars refused to begin their proj-ects from the dominant conceptual frameworks of their disciplines and institutions. The problem, they argued, was that such conceptual schemes had come to be highly valued and to seem rational only be-cause they articulated the concerns of men in the dominant social groups. This complicity with a commitment to male supremacy which such disciplines provided to dominant institutions could be detected only (or, at least, most easily) if one looked at the priorities of research disciplines through questions arising from women's everyday lives. Feminist work did so in order to reveal and challenge dominant institutional understandings of women, men, and social relations between them. It is worth repeating that while they began their projects by thinking about social relations through questions arising in women's lives, the point of such work was to "study up." From the beginning, their project was to reveal the principles and practices of dominant institutions, including research disciplines, which "governed" women's lives. Ethnographies of women's lives can be useful to standpoint projects. Yet standpoint projects insist on looking at the ways in which women's lives are enabled and constrained by the assumptions and practices of dominant institutions, including the research disciplines which serve such institutions and their public policy. Critical institutional ethnographies have become an important part of such work (Smith, *Institutional Ethnography*).

The logic of scientific inquiry "from below"

Articulating the "logic" of standpoint research in terms of such conventional scientific goals as good method and the objectivity of research enables the strengths of this approach to be grasped more clearly. Like a conventional philosophy of science, this is indeed a "rational reconstruction" of what successful researchers have done—in this case, feminist natural and social science researchers. Yet, in contrast to conventional philosophic practices, it is also intended to change future research practices—to change the way science is done. Standpoint theory, like the rest of feminist science studies, is intended as a desirable part of scientific practice. It is part of the "sciences of science" and is internal to the production of scientific knowledge. How is this kind of apparently illicit feminist research practice to be understood?[15]

Doing and knowing. How societies are structured has epistemological consequences. Knowledge and power are internally linked; they co-constitute and co-maintain each other. What people do—what kind of interactions they have in social relations and relations to the natural world—both enables and limits what they can know.[16] And what people typically can "do" depends in part upon their locations in social structures. Some are assigned the work of taking care of children and of people's bodies and the spaces they inhabit. Others are assigned to such jobs as administering governmental agencies, corporations, or research institutes; this is the form that "ruling" takes in our kinds of societies (cf. Smith, *Conceptual Practices of Power*). People's different experiences of interacting with nature, other people, and social institutions make some explanations of their own characteristics, actions, and the world around them look reasonable, and others unreasonable or even unintelligible. Thus material life both enables and limits what people can come to know about themselves and the worlds around them.

So the economic, political, and social structures of societies provide a kind of laboratory or field site within which we can observe how different kinds of assigned or chosen activities each enable some insights and block others. What does the person (female or male) who mothers a child understand that the doctor or lawyer probably does not on the basis of his/her experience, and vice versa? And we can learn from new experiences such as trying to live in sustainable relations to our natural environments, living with a two-year-old, or helping to organize an effective community council. Feminist biology, health, and environmental projects started off from women's experiences of their own bodies and those for whose care they were responsible as mothers, family health care providers, nurses, or community workers. Further research identified the sexist and androcentric metaphors, models, and analogies which shaped and often distorted such fields as physics, chemistry, engineering, and even statistics (see earlier citations.) Recollect that Beck and Gibbons, Nowotny, and Scott identified the emergence of everyday experience as an important new component of the production of scientific knowledge. Beck suggested this phenomenon be conceptualized as a science of questions.

Margins as sites of radical epistemological possibility. bell hooks ("Choosing the Margin") argues that margins are sites of potentially radical critical thought. Thus when material life is hierarchically organized, as in societies structured by class, gender, race, ethnic, religious, or other forms of oppression and discrimination, the understandings of such hierarchical relations that are available to "rulers" and "ruled" will tend to be opposed in important respects. The understandings available to the dominant group tend to support the legitimacy of its dominating position, while the understandings available to the dominated tend to delegitimate such domination (Hartsock, "Feminist Standpoint" 287). The slave-owner can see slaves' actions only as (unwilled) behavior caused by slaves' inferior nature or by their master's demand for obedience: he commands and they obey. Slaves don't appear to be fully human to their masters. However, following the slaves around in their everyday life, one could see their purportedly natural laziness as the only kind of political protest that they think they can get away with. One can see them smiling at the master as a subterfuge to obscure that they are secretly planning to run away or perhaps even to kill him. One can see them struggling to make their own human history in conditions not of their choosing (to paraphrase Marx). To take another example, Marx and Engels explained how the nineteenth-century capitalist economy worked from the perspective of workers' lives, contrary to the then-dominant understandings of it constructed from the experiences of the owners of industries and the financiers who served them.

Similarly, the women's movement of the 1970s revealed how women's work was both socially necessary and also exploited labor, not just an expression of women's natural inclinations or only a "labor of love," as men and public institutions saw the matter. Feminists pointed out that women never asked for or deserved rape or physical violence, contrary to the view of their abusers and the legal system. Rather, as MacKinnon ("Feminism, Marxism, Method") argued, "the state is male" in its insistence on regarding as objective and rational a perception of violence against women that could look reasonable only from the perspective of men's position in social relations between the genders in our particular kinds of societies. Again, biologists, health researchers, and environmentalists identified many more inversions and, from the standpoint of women's lives, perverse understandings of nature and social relations in the conceptual frameworks of dominant institutions. These frameworks are in fact active agents in forming and maintaining gendered social relations. As Fredric Jameson points out, feminist standpoint theorists opened "a space of a different kind for polemics about the epistemological priority of the experience of various groups or collectivities (most immediately, in this case, the experience of women as opposed to the experience of the industrial working class)" (Jameson, " 'History and Class Consciousness' " 144).

How the world comes to conform to dominant beliefs. Thus the oppressors' false and perverse perceptions are nevertheless made "real" and operative, for all are forced to live in social structures and institutions designed to serve the oppressors' understandings of self and society. These hierarchical structures and institutions engage in conceptual practices that solidify their continued power through disseminating such practices as natural, inevitable, and desirable. Conceptual practices have material effects. (Recollect how this point was developed in Beck and Gibbons, Nowotny, and Scott.) Social and natural sciences play an important role in developing and maintaining such ideologies, involuntarily or not.[17] If women are not permitted training in Latin, logic, science, computers, or public speaking (or whatever schools think important for the boys who will become societal leaders to learn), they will appear less rational than their brothers. If they are discouraged from physical exercise and sports, they will appear naturally weaker than and thus inferior to their brothers. If they are not permitted legal or philosophic training in favored forms of sound argument they will appear less capable of reasoned moral judgments in the eyes of legal systems and religious institutions. If they are encouraged to pitch their voices at the high end of their natural register and always to smile or look pleasant, while their brothers are encouraged to pitch their voices at the low end of their natural register and in public appearances to look serious, aggressive, or even on occasion angry, women speakers will appear less authoritative. Dominant social relations can make real many aspects of the worlds that they desire.

The need for science and politics. Consequently, it takes both science and politics to see the world "behind," "beneath," or "from outside" the oppressors' institutionalized vision. Of course no one's understanding can completely escape its historical moment; that was the positivist dream that standpoint approaches deny. All understanding is socially local, or situated.[18] The success of standpoint research requires only a degree of freedom from the dominant understanding, not complete freedom from it. And such freedom requires collective inquiry, discussion, and struggle for a marginalized group to "come to voice" as a self-consciously defined group *for* itself (instead of only an "objective" group in the eyes of others). Thus a standpoint is an achievement, not an ascription; and it is a group achievement, not something an individual can achieve apart from an emancipatory social movement or context. Women do not automatically have access to a standpoint of women or a feminist standpoint. Such a standpoint must be struggled for against the apparent realities made to appear natural and obvious by dominant institutions, and against the ongoing political disempowerment of oppressed groups.

Dominant groups do not want revealed either the falsity or the unjust political consequences of their material and conceptual practices. They usually do not know that their assumptions are false (that slaves are fully human, that poverty in any given society probably is not a consequence of high birth rates, that men are not the only reasonable or desirable model of the ideal human), and do not want to confront the claim that unjust political conditions are the consequence of their own views and practices. It takes "strong objectivity" methods to locate the practices of power that appear only in the apparently abstract, value-neutral conceptual frameworks favored by dominant social institutions and the disciplines that service them (to be discussed further below) (Harding, "After the Neutrality Ideal" and *Is Science Multicultural?*). Importantly, the standpoint claim is that these political struggles that are necessary to reveal such institutional and disciplinary practices are thus themselves systematically knowledge producing.[19] Politics and knowledge prove in principle no less than in practice impossible to separate since the very production of knowledge requires political action and has political effects. Effective political action requires reliable knowledge of the world and, in turn, produces it.

Knowledge for the oppressed and for a democratic social order. Finally, the achievement of a standpoint can empower an oppressed group.[20] An oppressed group must become a group "for itself," not just "in itself" —as others observe it—in order for it to see the importance of engaging in political and scientific struggles to see the world from the perspective of its own lives. Women have always been an identifiable category for social thought—an object conceptualized from outside the group— namely, from the perspective of men. But it took women's movements for women to recognize their shared interests and transform themselves into groups "for women," defining themselves, their lives, their needs, and their desires for themselves. Women's movements created a group consciousness (or, rather, many different group consciousnesses in different groups of women) in those who participated in them (and many who only watched) that enabled feminist struggles and then further feminist perceptions. Similarly, it took civil rights struggles and Black nationalist movements of the 1960s to mobilize African Americans into collective political actions that could, it was hoped, end racial inequities. The Chicano/a movement developed to mobilize Mexican Americans to a group consciousness capable of advancing an end to the injustices visited upon them. The Lesbian and Gay Pride movement had a similar goal and effect. New group consciousnesses were created through these processes, consciousnesses that could produce new understandings of social relations, past and present, and that could engage in political struggles for themselves.

Thus, starting off research from the lives of people in groups that are absent from the design and management of the institutions which administer everyone's lives has both scientific/epistemic and political consequences. How one actually goes about designing such research is and should be perpetually open to critical examination in particular cases. Yet it is by now easy to find models for such projects in many social and natural science fields, and no field is immune to such work.

Let us summarize the strengths of the feminist science studies and especially the standpoint approaches, identifying as we go along how they contribute to rethinking modernity. In the concluding section we identify some limitations of these studies.

4. GENDER, SCIENCE, MODERNITY: STRENGTHS

First and most obviously, these feminist accounts center women as agents, as subjects, of science and of history. They are interested in what the sciences do and could look like if one starts off thinking about them from women's lives. Of course they do feature scientific accounts about and by women. Yet their primary focus is on how starting off thought or research from women's lives gives us new and illuminating understandings of the sciences and how the latter work up the "buzzing, blooming confusion" of experiences of daily life into the kinds of categories and causal relations of social and scientific theories that are useful for the few to govern the many. While there are gestures to this work in Latour, Beck, and Gibbons, Nowotny, and Scott, those authors do not engage with it. It does not shape their own accounts and they do not recommend that it shape future such accounts. The feminist accounts alone take women to be fully human, as human as their brothers, and therefore also rightful subjects—that is, speakers, agents—of science, theory, and history.[21]

I cannot move past the preceding paragraph, however, without noting, again, that the internal logic of feminist analyses of gender—its necessary co-constitution with class, race, and whatever other social hierarchies shape a particular social context—in principle leads these science studies to consideration of the different historical groups of women who must be the subjects of feminist knowledge. The subjects of feminist science studies are and must be plural. I say "in principle," because while the intersectionality of gender is brilliantly conceptualized and researched in the work of some of the scholars in this field, it tends to get lost in much of the rest of the actual research.[22] Of course, this is not the result of overtly racialized or imperial mentalities on the part of the researchers. Rather their feminist research projects are situated also in mainstream history, sociology, and philosophy of science. These fields have been preoccupied with the "high sciences" of mainly physics or microbiology and with the laboratory practices, which have been overly centered in social studies of sciences in the North. This large part of feminist science studies is situated within exceptionalist and triumphalist accounts of Northern sciences and their philosophies, and it retains those exceptionalist and triumphalist horizons characteristic of modernity. It is functionally racialized and imperial, not intentionally so. We return to this point in the next chapter.

Next, these studies argue that political engagement is necessary to transform the sciences; the epistemological core of the sciences cannot be isolated and revised apart from its political contexts. Moreover, it is not feminisms which introduce politics to sciences and philosophies of science that were otherwise "pure" and politically neutral. Rather feminisms identify the gender politics which can and often do shape every stage of scientific research, including its cognitive, technical core and its standards for adequacy. To put the point another way, mainstream science is functionally male-supremacist, whatever the intentions of its creators. It functions to advance male-supremacist research, knowledge claims, and their epistemologies and philosophies of science. These feminist science studies intend to balance—or, rather, to transform—the existing politics of the world of the sciences into ones that recognize women as just as paradigmatically human as their brothers.

It is indeed women's social movements which are necessary and have acted to transform the sciences and politics of our worlds. Politics is not just a matter of individuals' political preferences. Rather, in contexts of oppression and exploitation, it is a matter of disadvantaged peoples organizing to become groups "for themselves," not just as objects of others' observations, as groups "in themselves." Women must come to see that it is not just "their man"—father, husband, colleague, or boss—who is "mean" to them. Rather, the dominant institutions of society, their practices, and their cultures have been organized to extract labor and caring from women and donate their value to those institutions and men in the dominant groups. This focus on the power and value of movements for democratic social transformation is centered in the standpoint projects where such social movement politics are in effect part of feminist research methods. It is present in a lesser but important way in the feminist empiricist accounts where such movements make visible to everyone certain kinds of social phenomena which were otherwise invisible (cf. Millman and Kanter, Introduction to *Another Voice*).[23]

Thus these studies expand what counts as the realm of the political in two ways. Along with the rest of the diverse feminist movements around the globe, they identify the politics of gender—"sexual politics" —which, in this case, shapes the institutions, practices, and cultures of the sciences. They also identify the politics of the sciences which center only the experiences, interests, values, standards, and dreams of men in the dominant social groups. In both cases they identify realms of the political expanded beyond the governmental realms in which modern Liberal politics locate political events and processes.[24]

Feminist science studies intends to be part of the culture of the sciences themselves rather than only an external criticism or analysis of it, in contrast to the way both conventional philosophies of science and the sociology of scientific knowledge conceptualize their projects. Thus FSS intends to change scientific practice itself and to contribute to the transformation of the sciences. This work distances itself from the conventional position philosophers of science take as only "handmaids" to the glorious achievements of the great figures in the history of Northern sciences. It also distances itself from the purely descriptive aim of many science studies sociologists, ethnographers, and historians. For the most part, the sociology of scientific knowledge, too, has no intention of changing how science is done but only how we understand it and its relation to its historical social eras. The three theorists discussed in Part I theoretically would include feminist science studies in the "sciences of science" they recommend.[25] But in actual practice they do not so include them.

Thus the feminist accounts expand the sciences in three ways. First, they produce "sciences of sciences," as Latour and Beck put the point, as does the rest of the field of postpositivist science studies. They produce accounts of how scientific research works which do not necessarily match the perceptions of the scientists who do the research, let alone the ideology of science defended by science policymakers when requesting funds from industry or government or trying to protect the sciences from, as they see it, excessive governmental or public oversight. Second, in the hands of standpoint projects, these sciences are to be self-critical in a stronger way; they are to exercise a "robust reflexivity" in which they invoke the same principles and standards in criticisms of the foundations and guiding principles of their own projects that they recommend for studying the rest of nature and social relations. They do not regard their own work as an exception to the scientific principles they otherwise advocate (Harding, *Is Science Multicultural?*, chap. 11; cf. Elam and Juhlin, "When Harry Met Sandra"). As indicated, there are unfortunate limitations on the critical perspectives which most of them actually do consider. My point here is that they promote a logic of critical, scientific studies of sciences including their own, and of their institutions, practices (including knowledge claims), and cultures.

In the third place, they provide systematic empirical accounts of gender politics, in this case in the institutions, practices, and cultures of the natural and social sciences. In both cases they expand the sciences to include everyday experiences and public discussions which produce "sciences of questions," as Beck put it. In this case, these "sciences of questions" emerge from women's movement discussions of women's experiences and of the scientific disciplines' "conceptual practices of power" (Smith, *Conceptual Practices of Power*).

Another strength is that feminist standpoint theory holds that how people live together—their distinctive social relations with each other and their interactions in socially typical ways with the natural world—shapes what they can know about themselves and the world around them. Thus this work implies that if we would transform the sciences, we must also transform the larger social relations that end up giving content, form, and value to existing kinds of scientific inquiry. We must learn to live together around the globe, in our governmental units, in our local communities, in our workplaces, and in our households, and to do so in geopolitical organizational contexts and with kinds of political, psychic, and cultural commitments very different from those experienced even by our parents. Of course this is not simply an abstract point for feminist movements; feminisms have produced volumes about how such social relations should change.

Clearly these feminist science studies have deeply challenged central features of modernity. They challenge the value-neutrality and progressiveness of the sciences, their rationality, and their philosophies. As part of their work within the culture of the sciences, feminist science studies have not only criticized the inadequate conceptions of rationality, objectivity, good method, and good science which have guided the existing sexist and androcentric research of the natural and social sciences. They have also produced radically revised standards for the adequacy of scientific research.

These studies show how men's distinctive gender fantasies haunt the way they think about public life and about the sciences and their rationality which are supposed to direct it. Where do these gender fantasies come from? In chapter 8 we will see accounts of their origins in men's early experiences of family life and in their economic and political interests in the family as well as at work, as these are shaped in adult social relations in modern societies. Thus these accounts challenge the division between public and private realms insofar as they show how gender relations in "private life" come to shape the content, standards, and visions of sciences. They challenge the purported autonomy of the sciences from other social institutions, such as politics and the family; yet this autonomy was supposed to be a central feature of modernity.

However, these feminist science accounts do not reject modernity. They have had an ambivalent relation to the field of postmodernism. In significant ways they are fully part of the modern valuing of criticism, the growth of knowledge, and more perfect sciences. They want expanded democratic principles and practices, not even more desiccated ones.

5. LIMITATIONS

Four limitations of these accounts are already visible. First, there is the matter of Eurocentrism, as I have been pointing out. In principle, feminism's intersectionality thesis should routinely direct self-critical assessments of overgeneralizing from one's own life to the lives of women living in different conditions. Yet in practice, most Northern feminist science studies—with the kinds of brilliant exceptions indicated earlier—have continued to be contained by the standard Northern Eurocentrism and its class commitments. It has had little interest in how Northern sciences and their philosophies emerged from long histories of imperialism and colonialism, or in exploring the indigenous science and technology projects of non-Western peoples. Nor has it often looked at Northern science theories and practices from the standpoint of the many kinds of women who do in fact work in or on behalf of scientific institutions in positions other than those of fully credentialed scientists: women lab technicians, women science students, male scientists' wives, or women science writers. Consequently, though this work has brilliantly challenged Western modernity from within such confines, it has nevertheless used the modern mostly as a horizon. It has rarely explicitly critically engaged with the binary of modern vs. pre-modern or tradition. It rarely thinks about Western sciences starting off from the postcolonial science studies grounded in Third World peoples' experiences, which will be the topic of the next chapter. Consequently, this work functionally retains powerful elements of the exceptionalist and triumphalist stances characteristic of modern philosophies of science and of mainstream modernization theories. This is so regardless of the intentions of its authors.

Thus one must ask if the conceptual frameworks which have proved so fruitful in Northern feminist challenges to modern sciences, their practices, and their cultures also prove to be the most fruitful for women in the South? Probably not—partly for the kinds of reasons already indicated. For example, philosopher Uma Narayan ("Project of a Feminist Epistemology") points to the fact that positivism has never been the major epistemological problem facing Southern women's criticisms of scientific practices—European or indigenous—in their own societies. Moreover, Northern feminist conceptions of ideal feminisms, of women in the South, and of other central feminist themes frequently reveal Eurocentrisms of their own (cf. Mohanty, *Feminism Without Borders*; Oyewumi, *Invention of Women*; Spivak, "Can the Subaltern Speak?"). And we will see that persistent exceptionalist assumptions in the Northern accounts will conflict with the postcolonial histories and social maps on which Southern feminist science studies are set.

In the second place, this work certainly understands that to make the sciences more democratic one must transform their surrounding politics also, and it is guided by women's movement discussions of what a non-sexist, non-androcentric society might look like. Yet it has not tended to locate its analyses on a map of progressive social theory or political philosophy more generally. It appears to think that work is to be done by sociologists and political philosophers. Of course this distance between science studies, on the one hand, and social theory and political philosophy on the other hand, is characteristic not just of the positivist sciences and their philosophies, but also of most of the field of Northern philosophy and social studies of science. Yet this is unfortunate, for the social theorists and political philosophers also tend to be leery of thinking simultaneously about science and society. If politics permeates the sciences, and if those sciences then come to permeate modern democratic politics, clearly the two must be transformed together, as Latour, Beck, and Gibbons, Nowotny, and Scott argued, in different ways.

One problem here is that Northern feminist science studies researchers and scholars tend to conceptualize the modernity of Northern sciences exclusively in abstract philosophic terms. They do not think of it also sociologically and historically, for example, as modernity appears in the late-nineteenth-century theory of modernization intended to explain the industrialization and urbanization emerging in Europe and North America at the time. Nor do they think of it as a resurgent defense of post–World War II, European/American economic flourishing and the Northern modernization projects for Third World so-called development.[26]

As we will see in the next few chapters, this is not a criticism that can be raised against the Southern feminist science and technology studies —about the feminist work set in the context of postcolonial science studies. Feminists in the South always must think about the global political economy and how it affects the lives of women in their societies. They do have some clear ideas about how society and sciences must be conjointly transformed. Indeed, when women in the North do start off thinking about sciences and technologies from the standpoint of women in the South, with the assistance of the critiques of Northern imperialism and of modernity one can find in such work, their focus on sciences in the North immediately becomes framed by "intersectional" transformative social theory and political philosophy. It becomes part of "Southern feminist science studies!"[27]

A third concern is that there is one issue of social theory introduced by Beck which is hard to find engaged in any of these feminist studies, North or South. This is the issue of constructing theories and practices of science which take into account the impossibility of knowing what the future will bring, the loss of a rhetoric of scientific certainty, and, most importantly, the probability that feminist work itself will change the world in unpredictable ways. With their focus on multiple sciences, the postcolonial science studies and feminist work within it discussed in the next two chapters get closer to appreciating this anxiety-producing fact of contemporary life.

Finally, what about the "counter-modernities" generated by both modernity and even the most progressive attempted transformations of it identified by Beck? Where has feminist science studies engaged with the challenge of incorporating from the start recruitment of counter-modern forces into progressive alliances?

5.

POSTCOLONIAL SCIENCE

AND TECHNOLOGY STUDIES

Are There Multiple Sciences?

— — — —

THE LAST QUARTER OF THE TWENTIETH CENTURY produced three
distinctive fields of social studies of science and technology. Post-
Kuhnian science and technology studies and feminist science and technology studies both began to develop from a few probing questions into
full-fledged complex intellectual and social movements. Meanwhile,
postcolonial science and technology studies (PCSTS) was also gathering
steam. It was fueled by the desire to tell a counter-version of the histories and present practices of both non-European and European
sciences, and especially of the long history of interactions between
them. These scientists, scholars, and activists also wanted to provide a
counter-narrative to the triumphalist Western account of Third World
development policies. In the triumphalist narrative, transfer of Western sciences and technologies and their rationality to the "underdeveloped societies" would bring social progress to the Third World.[1]

In its early days, PCSTS worked for the most part under the radar of the
Northern-based science and technology studies movements. Recently
this situation has been improving. By the 1980s, some of this work
began to become available in English. Northern-trained historians and

philosophers of science began to join their Southern colleagues in producing it. International agencies such as UNESCO began to support conferences and publications on PCSTS topics (Gender Working Group, *Missing Links*; Goonatilake, *Aborted Discovery*; Harding and McGregor, "Gender Dimension"; Moraze, *Science and the Factors of Inequality*; Nandy, *The Intimate Enemy*; Sardar, *The Revenge of Athena*; Waast, *Sciences in the South*). Today one could justifiably say that the field has begun to flourish in the North, too. Illuminating studies have been appearing with increasing regularity; sessions of international conferences in the North are beginning to be organized around engagements with PCSTS; leading academic programs in the United States are seeking faculty to teach in these areas; graduate students are choosing to work on such topics; and there is increased awareness of at least a few of the issues in popular media. This is not to say that the conceptual legacies of imperialism, with their exceptionalist and triumphalist visions of Western Civilization and its sciences, have disappeared (Willinsky, *Learning to Divide the World*; Hobson, *Eastern Origins of Western Civilisation*). By no means is this the case. Yet political and conceptual resources have been gathering on behalf of PCSTS concerns. It is becoming increasingly clear that what constitutes social progress for the South must be defined within those societies. The debates over how Northern and Southern science and technology traditions and projects contribute to social progress in the South, and what the relationship between them has been, is now, and should be—these are controversial issues in the South, too, and will remain so for many decades to come.

How do the PCSTS issues differ from those of the Northern accounts? In what ways are the exceptionalism and triumphalism of Eurocentric science studies rejected in these postcolonial studies? What are their arguments for multiple sciences, each with its own cultural legacy? How are visions of and plans for sciences and democratic social and political practices linked here? How are Northern understandings of modernity and of tradition challenged in PCSTS? Issues about gender emerge within all of the PCSTS concerns. Those issues are reserved for the next chapter.

1. ORIGINS

The origins of PCSTS can be found in at least four places. One was in historians' and geographers' questions about causal relations between

European expansion ("The Voyages of Discovery") and the emergence of modern sciences in Europe at about the same time. Was it possible that these two significant events in European history were not at all causally linked, as the standard accounts implied? Related to this work is a recent tendency with older origins.

Here historians ask about the processes through which Asian scientific and technological advances reached Europe a millennium and more ago, and how and why the origins of these borrowings were then overtly repressed (Lach, *Asia in the Making of Europe*; Needham, *Science and Civilisation in China*; Sabra, "The Scientific Enterprise"). Indeed, Hobson recently has argued that there was little of value to Asian cultures in European culture and society, including its sciences and technologies, until perhaps as recently as the late eighteenth century and early nineteenth. He proposes that the suppression of European borrowings from Asia is at least in part explained by how early European identity was formed explicitly against Islam—and, indeed, continues to be so defined today (Hobson, *Eastern Origins of Western Civilisation*). One distinctive feature in this kind of study is how it reverses the imperial gaze. Hobson consistently organizes his account through a gaze—a standpoint—that travels from East to West rather than in the reverse direction which has characterized Eurocentric studies. This practice can intermittently be found in the earlier works cited.

A second source used an anti-Eurocentric lens to reexamine scientific and technological traditions of non-European cultures. This project was developed in the newly independent ex-colonies, as Eurocentric narratives of "savage minds" and universally valid "international science" were replaced by more balanced studies of the economic and political roots of Northern sciences and of the history of empirical knowledge-seeking in their own societies. A number of scholars of Northern origin also produced such accounts beginning in the 1970s.

A third source was in anthropologists' recognition, beginning in the late 1950s, that they could use "reflexively" their distinctive research methodologies which had been developed to understand the ways of life of other cultures. They could use these to study cultural features of Northern scientific and technological practices.[2] Latour's and Woolgar's study of laboratory life (*Laboratory Life*) and Sharon Traweek's (*Beamtimes*) comparative study of Japanese and European/U.S. high-energy physics subsequently provided influential models for this kind of work.

A fourth origin produced criticisms of the imperial and neocolonial character of the Northern development policies for the Third World, which were initiated in the 1950s. From its start, so-called development was conceptualized as the transfer of First World scientific rationality and technical expertise to the underdeveloped societies, as they were then called. Thus the practices and philosophies of First World sciences and technologies would be implicated in the successes and, as it turned out, the failures of Third World development (cf. Escobar, *Encountering Development*; Sachs, *Development Dictionary*). "Modernization" and "globalization" today continue European expansion by another name, aided by scientific and technological resources.[3]

These postcolonial science and technology studies emerged as a distinctive field—or, rather, several related but not-always-communicating fields—by the early 1980s. They have now accumulated more than two decades of books, articles, journals, conferences, manifestos, websites, and a significant presence in ongoing projects of United Nations organizations as well as other national and international institutions and agencies. Issues about them appear in the indigenous property rights controversies, in how "the nature of science" should be taught in the K-12 grades, and in a number of other vibrant discussions going on around the world. There has always been at least a sprinkling of Western activists and scholars involved in these projects, and many more so in recent years.[4]

Yet preoccupation in Northern science and technology studies with microstudies of Northern labs and field sites, and with the history of the achievements of Northerners and especially of the Great Men, has slowed Northern engagement with PCSTS. I do not mean to disvalue this immensely illuminating work of the sociologists, ethnographers, and historians. It has helped to level the playing field between Western and non-Western sciences by demonstrating how deeply embedded in historical social projects even the greatest achievements of Western sciences, not just their most obviously erroneous claims, have been. This work undermines empirical support for the conventional contrast between purportedly value-free Northern science and value-laden knowledge systems of other cultures, though this consequence is not much acknowledged in Northern science and technology studies. Rather, my point is that its fundamental assumptions make it difficult to think other cultures' science and technology traditions worth engaging. These frameworks obscure the value of taking a postcolonial standpoint. Moreover, it is by no means only a phenomenon of the Eurocentric past to encounter among educated Northerners even the overt assessment of other cultures' knowledge systems as only superstition and myth.[5] I stress "overt" to point not only to wide-spread Eurocentric ignorance, but also to the disturbing comfort such teachers, scientists, and scholars still feel in dismissing the possibility of noteworthy scientific and technological achievements of non-European cultures. The Western imperial legacy is still entrenched in institutions of higher education no less than in education at lower levels (Willinsky, *Learning to Divide the World*).

This chapter identifies central themes in this work and ways in which it also proposes to transform deeply intertwined scientific and political philosophies and agendas. But first a reminder about what is involved in taking the standpoints of peoples *not* of European descent, wherever they live, who have borne most of the costs and received fewest of the benefits of Northern sciences and technologies.

2. STANDPOINTS OF SOUTHERNERS

PCSTS take the standpoint(s) of non-European cultures in order to reexamine critically both Northern and Southern scientific and technological traditions. These accounts start off with issues arising in the lives of people in those cultures, past and present, instead of from the familiar conceptual frameworks of research disciplines. Then they evaluate critically the policies and practices of the North which have created such issues in different ways in different cultures. Thus these projects "study up." And they can also then reevaluate indigenous knowledge and traditional environmental knowledge not from the perspective of conventional Northern exceptionalist and triumphalist standards, but rather as projects which responded well, or not, to concerns of non-European societies and their peoples. PCSTS present a counter-narrative to the standard Northern exceptionalist and triumphalist narratives of human scientific and technological achievements.

There are several ways in which this kind of standpoint work can be done.[6] One can begin with the experiences and voices of the peoples Europeans encountered, past and present. Those of the past are not always easy to locate. Yet this challenge is no harder for science and technology studies than for researchers who seek such accounts from any other "silent" groups in history—peasants, women, subalterns (cf. Spivak, "Can the Subaltern Speak?"; Bridenthal and Koonz, *Becoming Visible*; Wolf, *Europe and the People Without a History*). What have been the science and technology concerns of these peoples? What are they today? Second, one can begin with the objective location in local and global political economies of these peoples, as evaluated through studies of economic production, trade, migration, global financial policies and flows, international political relations, international development policies, comparative census data, global health statistics, and other such measures. World Systems Theory and Dependency Theory have provided powerful resources here (Frank, *Capitalism and Underdevelopment in Latin America*; Wallerstein, *Modern World System*).[7] A third kind of resource has been found in the collective statements, analyses, manifestos, calls to action, protests against local or global policies, and so on of non-European peoples (for example, Third World Network, *Modern Science in Crisis*). All three kinds of sources are revealing, and no one is in itself sufficient to fully examine or understand science and technology histories and present practices around the globe. The larger field of postcolonial studies has addressed these kinds of issues in great detail, though with little attention to science and technology (e.g., Ashcroft, Griffiths, and Tiffin, *Postcolonial Studies Reader*; Williams and Chrisman, *Colonial Discourse*).

These standpoints enable both the constituencies that postcolonialism represents and also the rest of us to understand aspects of Northern and Southern modernities and their sciences which were invisible or at least not easily detected from the typical perspective of activities of economically and politically privileged Northerners. These are the ones who have received most of the benefits and fewest of the costs of Northern modernization and its scientific and technological transformations. Such postcolonial standpoints enable us all to avoid restricting ourselves to only the "Northern native's" view of our own and others' practices and cultures. After all, Enlightenment assumptions constitute many everyday beliefs and practices of Northerners. Thus we, too, can achieve more balanced, objective accounts of how modernization and science and technology policy have operated in the past and continue to do so today. Such projects enact a "robust reflexivity" which insists on applying the same scientific/critical standards to our views of ourselves and the world with which we interact that we recommend to others. Such standpoints move beyond the important but too limited attention to "inclusive" anti-Eurocentric gestures and practices (cf. Harding "Robust Reflexivity"; Elam and Juhlin, "When Harry Met Sandra").[8]

Science and empires studies

Science and empire studies emerged alongside World Systems Theory. Eric Williams (*Capitalism and Slavery*), a West Indian historian, looked at how the immense profits from Caribbean plantations had played such a large role in making industrialization in Europe possible. Several decades later, Ramkrishna Mukerjee (*Rise and Fall*), an Indian historian, began to examine how the British intentionally destroyed the Indian textile industry in order to create a market for the importation of British textiles.[9] Scientific and technological knowledge, both in Europe and in Europe's overseas targets of imperial control, were central to both of these histories. In what was perceived then as "externalist history," it began to appear that European expansion and the destruction of other cultures' knowledge traditions bore significant responsibility for scientific and technological growth in Europe. Thus doubts arose about the prevailing diffusionist model of scientific and technological growth, and the presumed political innocence of Europe in such processes.[10]

Science and empires scholarship has radically expanded since these early works. These scholars ask if it was entirely an accident, as the standard histories of Northern science assume, that modern sciences began to flourish in Europe at about the same time as the Europeans began their "voyages of discovery." Their answer is that it was no coincidence. Rather, each project needed the success of the other for its own success. Moreover, this symbiotic relation between European expansion and the advance of modern sciences continues today through development projects and Western militarism (especially, it is mortifying to admit, U.S. militarism)—a point to which we return shortly.

The three great corporate sponsors of the voyages were the Jesuits, seeking to create Christian souls; the great European trading companies, such as the Dutch and British East India Companies; and the imperial and colonizing European nations seeking prestige, power, and the gold and silver, furs, spices, and other riches they hoped their voyages could bring back to Europe (Harris, "Long-Distance Corporations"). The Europeans needed the kind of information about nature's order necessary to establish global trade routes, plantations, and settlements in the Americas and, eventually, in Australia, Asia, and Africa. They needed advances in navigation and thus in the astronomy of the Southern Hemisphere, in oceanography and climatology, in cartography, botany, agricultural sciences, geology, medicine, pharmacology, weaponry, and other fields that could provide information enabling Europeans to travel far beyond the boundaries of Europe and to survive encounters with unfamiliar oceans, lands, climates, flora, fauna, and peoples, as well as with their European competitors (cf. Blaut, *Colonizer's Model of the World*; Crosby, *Columbian Exchange*; *Ecological Imperialism*).

But the production of such information in turn required European expansion. Europeans foraged in other cultures' knowledge systems, absorbing into their own sciences useful information about the new environments they encountered, as well as new research technologies and conceptual frameworks used in other cultures. Moreover, before the European voyages, other cultures had established pluricentric trading routes which gave each access to aspects of nature different from those in their homelands. European expansion reorganized travel around the world so that only, or at least primarily, Europe became the center of global trading routes. Thus Europeans' appropriation of access to nature around the globe enabled them also to compare, contrast, and combine observations of nature's regularities in different geographical sites. (Consider, for example, how important such access was to Darwin several centuries later.) Furthermore, through expansion, potentially sophisticated competitors to European science were accidently vanquished, for example, through the introduction of infectious diseases to which the indigenes had no resistance. And they were intentionally destroyed, as in the case of the Indian and African textile industries, along with the curtailment of the ability of such cultures to flourish. Sometimes the Europeans succeeded, intentionally or not, in wiping out the very existence of other cultures and their peoples. Thus the voyages of discovery and subsequent European expansionist projects greatly contributed to the way modern sciences flourished specifically in Europe, rather than also, or instead of, in other cultures of the day.

Writing of the two-century British occupation of India, historian R. K. Kochhar says that during this period India became a laboratory for British science ("Science in British India"). His point can be expanded to the European imperial and colonial projects more generally, as many scholarly studies reveal how through such projects the world became a laboratory for European sciences.[11]

These science and empire accounts reveal the Eurocentrism of standard histories, sociologies, and philosophies of science. They show the dependence of the development of Northern sciences upon Northern imperialism and colonialism. These voyages turn out to be one of the causes of the development of sciences in Europe and the decline of other cultures' empirical knowledge systems. Indeed, the conventional internalist "logic of scientific inquiry" and its related philosophic claims should be regarded in terms of their function as well as their intent. Whatever their intent, they function as a product and as a defense of European expansion. Of course, the great achievements of modern Western sciences are partly a consequence of distinctive "internal" processes: experimental method, a valuable ontology of primary and secondary qualities, a critical attitude toward received belief, and so forth. Yet they are also a consequence of material historical circumstances. Without such circumstances the achievements could not have occurred. The circumstances shaped not only the fact of the achievements, but also their content. This is not to blame individual scientists, philosophers, or historians for the invention and maintenance of this internalist view of the growth and flourishing of Northern sciences prior to the availability of the PCSTS accounts. They observed the world and thought about it within the kinds of sightings available to their time and place and to their positions in global political, economic, and social relations. Rather, the point here is to ask *us, today*, to provide a more objective account, with the hindsight that PCSTS help to provide, of the growth and development of Northern sciences and the achievements and often decline of the sciences of the societies the Europeans encountered. Our time and place are no longer the same as those in which conventional Western historians and philosophers could reasonably feel at home; if one feels comfortable with the conventional accounts in today's world, that is an intellectual and political problem.

Rethinking their "ethnosciences"

Alongside the science and empires studies, anthropologists and historians began to reevaluate and examine more thoroughly both the traditional environmental knowledge (TEK) of other societies and their indigenous knowledge (IK) traditions more generally.

Of course every society must have "the scientific impulse" in order to survive, as anthropologist Bronislaw Malinowski put it (*Magic*).[12] In the last few decades, a global "comparative ethnoscience" movement has flourished, stimulated by national/ethnic pride, the desire to preserve a disappearing part of the human scientific legacy, and the need to protect the knowledge systems of "pre-modern cultures" from looting by Northern pharmaceutical companies and other kinds of corporations seeking economic profits. International governmental agencies as well as NGOs have been active in sponsoring these studies of the knowledge systems of other cultures.[13]

One can characterize the resources of these universal scientific impulses with greater specificity.[14] Different cultures occupy and travel through different locations in nature's heterogeneous order: some live on the borders of the Atlantic Ocean, some on prairies; some live in the Arctic and others at the Equator. Even in the same location, different cultures can have different interests in their part of nature: some will want to use the Atlantic Ocean for fishing, others as a coastal trading route or a military highway, and yet others for mining the oil and gas under the ocean floor. Even a single cultural tradition can at times bring different discourses—narratives, metaphors, models, and analogies—to conceptualize the world around them: Northerners, for example, have conceptualized the planet as the Judeo-Christian God's gift to his people, as a cornucopia of limitless abundance for human use ("Mother Earth"), as a mechanism of one kind or another, and, thanks to recent environmental movements, as a spaceship or lifeboat whose occupants and their "vessel" must be carefully managed. Each metaphor or model directs different inquiry projects.

Moreover, cultures tend to produce knowledge in ways similar to how they produce other artifacts. They use "craft" or "factory" work processes, and they store and share the knowledge gained in ways similar to how they store and share other kinds of human products, including the legal control of such processes and their output in ways they do (or don't) control other kinds of valuable property. Finally, a culture's location in regional and global economic and political hierarchies will shape the way each of these aspects of their scientific impulses will be articulated. Rich and powerful cultures already own much of the "nature" and the productive mechanisms upon which such processes depend, and they can control access to the results of their work to a much greater extent than can poor and relatively powerless cultures.

TEK and IK work is important because it recognizes the high scientific and technological achievements of other cultures, and that in many cases these achievements occurred earlier than in the North. And it does so for peoples Europe has characterized in a derogatory fashion, sometimes during many centuries of encounters. Moreover, it identifies features of Northern sciences, past and present, which were learned from other cultures and usually unacknowledged in the standard histories of science. Additionally, it leads to questions about the North's unintentional and intentional destruction of these other knowledge systems and the ways in which the conventional internalist and triumphalist philosophies and histories of science function to justify both destroying and ignoring these achievements of other cultures. Finally, it draws attention to the systematic ignorance in the North about both the achievements of other cultures and the historical causes of, and failures and gaps in, Northern understanding of nature and social relations. This work contains some hard and painful truths for peoples of the North.

It also gives support to a related project, namely, conceptualizing Northern sciences and technologies as also local knowledge systems. This project has long been engaged by Northern science and technology studies, including their feminist components. But in the context of PCSTS, the "local" of interest centers Northern imperial and colonial activities.

Rethinking ours: Northern ethnosciences

The ethnographic study of Northern sciences was called for in a presidential address to the Anthropological Society of Washington in the late 1950s.[15] As we saw first in the discussion of Bruno Latour's work in chapter 1, this project leveled the playing field of comparative science studies by showing that whatever might distinguish modern sciences of the Global North from the empirical knowledge projects of other cultures, it certainly was not that the Northern ones alone managed to escape the cultures which produced and maintained them, and so become culture-neutral. Rather, these ethnographic studies showed that Northern sciences, too, are "socially situated," taking their problems, ontologies and other background assumptions, and preferred methods of collecting and evaluating data from their local historical contexts. To be sure, modern Northern sciences frequently challenged their own cultures' norms, but which ones they challenged and the ways in which they did so were themselves the product of particular historical eras. Northern sciences might travel further to become "international sciences," but that did not mean that they were value-neutral or culture-free, as we will explore further in chapters 7 and 8. Therefore, paradoxically from conventional perspectives, it had to be that knowledge claims could be universally useful and accepted without being culture-free. This work showed how Northern sciences, the most acclaimed achievements of the Enlightenment legacy, were nevertheless in distinctive ways still "ethnosciences." This focus of the comparative ethnoscience movement gave increased legitimacy to the other project of reevaluating the knowledge traditions of Southern cultures.

Of course the focus of the Northern science and technology studies was on studies showing how "socially situated" Northern sciences were and are, beginning with the work of Thomas Kuhn and the new social histories which inspired and informed his own study (*Structure of Scientific Revolutions*). For philosophers, the beginnings of this insight date back to W. V. O. Quine's criticisms of empiricism and his discussions of the continuities between science and common sense and how they are linked through "networks of belief" ("Two Dogmas of Empiricism" and *Word and Object*).

Another critical focus on Northern histories and philosophies of science looks at the typically inflated value Westerners place on their own scientific and technological legacies in world history. The Islamic world and China, India, Southeast Asia, and Japan all achieved periods of "high culture" long before modern sciences and technologies developed in Europe and even, in some cases, before Europe existed as an identifiable piece of geography or as a political and cultural identity. Islam extended deeply into Africa and Europe. These cultures' scientific and technological traditions were mature and sophisticated, as some historians long had recognized (Needham, *Science and Civilisation in China*; Sabra, "Scientific Enterprise"; cf. Goonatilake, *Aborted Discovery*; Selin, *Encyclopedia of the History of Science*).

An interesting turn in this literature appears in John Hobson's study of the "Oriental West" (*Eastern Origins of Western Civilisation*) and in his focus on the relatively neglected topic of the formation of European identity. He draws on a wide array of sources to pinpoint just how much the emergence and subsequent flourishing of modern sciences in Europe is indebted to Eastern cultures. He also traces the processes through which a motley collection of peoples became unified as a

European "Christendom." These included indigenous peoples living in the area, immigrants from various parts of Asia, and the internally displaced populations created by the conflicts between immigrants and indigenes. Hobson argues that the "threat of Islam" was intentionally used to constitute this distinctive European and Christian identity. It enabled the grossly inequitable European feudal system to continue with relatively little disruption from the exploited serfs through a reconceptualization of who the feudal classes were that was attractive to serfs as well as aristocrats. The exploited labors of the serfs, no less than the labors of their aristocratic rulers and the warrior knights, made crucial contributions to Christendom, on this account. The history of Islamic and Asian achievement of every sort was intentionally suppressed as part of the formation of European identity as superior to that of the "infidels."[16]

The legacy of these processes persists today, Hobson points out, in the widespread Eurocentric ignorance about the scientific and technological achievements of Asian cultures, their effects on the emergence of modern sciences and technologies, and the origins of European identity. One study of the active perpetuation of an imperial mindset can be found in Willinsky's account of Canadian education today (*Learning to Divide the World*). He shows in detail how the legacy of imperialism in North America systematically leaves children still learning "how to divide the world" in the ways desired by imperialism (British in this case).

Criticisms of Third World development policies

Criticism of the North's so-called development policies for the Third World emerged alongside those policies in the 1950s and 1960s. Development was conceptualized as modernization. This was the heyday of exceptionalist and triumphalist modernization theory among Northern social theorists. The Northern development agencies justified by appeals to humanitarianism their plans to bring the so-called underdeveloped societies up to the standard of living of the industrialized North, thereby ending the poverty which was seen as a consequence of the traditionalism of Southern societies.

From its beginnings, development was conceptualized as achievable only through the transfer to the South of Northern scientific rationality and technical expertise and the democratic political forms that these purportedly bring into existence and are, in turn, supported by (see, e.g., Snow, *Two Cultures*). Yet the policies responding to such appeals have not had the effects they promised. The gap between the rich and the poor has increased throughout this period in the South as well as in the North. It turns out that development policies, intentionally or not, largely continued the earlier imperial and colonial pattern of directing the flow of natural, human, and other economic resources from the South to the North, and from the least to the most already-advantaged groups within societies around the globe. The consequences have been maldevelopment and de-development for the majority of the world's peoples who were already the most impoverished. Meanwhile, we can now see that it is the investing classes in the North and their allies in the South who have in fact been developed by these policies as, by now, even the mainstream Northern press regularly reports. From South to North have continued to flow, first of all, money and other financial resources, mostly through the interest payments from the South for the development loans they received from Northern banks. But also joining this northward flow have been natural resources, "brain power" (as in "brain drains"), and cheap labor for domestic and service work, as well as manufacturing, agricultural and construction work, and access to TEK and IK traditions and their products. Moreover, it turns out that it is the North that over-reproduces. Poor people who make up the vast majority of citizens in the South (and the North) cannot reproduce themselves—contrary to the popular assumption (cf. Escobar *Encountering Development*; Sachs, *Development Dictionary*).

What would development for poor peoples in the South and North look like if it were directed by the interests and desires of poor peoples themselves instead of by the economic and political agendas of elites in the North and their allies in the South? What would contemporary Southern sciences and technologies look like if they could flourish in such a Southern-directed context rather than in the doubly dangerous environment of, on the one hand, development agencies' traditional romanticized views of Northern sciences and technologies and, on the other hand, the kinds of resistance to the latter that can still appear on behalf of romanticized traditional knowledge systems of the South?

3. POSTCOLONIAL PROJECTS

This postcolonial work raises important issues neglected by Northern science studies, including Northern feminist work. Some of the most significant are the following.

(1) Inclusion and beyond

Just including accounts of the scientific and technological traditions of other cultures, but without stigmatizing them, in standard histories of science clearly would be an important step. Indeed it could be considered radical to do so without disvaluing them as primitive, merely magic and superstition, merely technological achievements, or in any of the other ways in which Northerners have maintained an exceptionalist history and philosophy of Northern sciences. A vast revision in world history is only now getting under way. Including accounts informed by postcolonial scholarship of other societies' scientific and technological achievements makes important contributions to this project.

Yet PCSTS make clear that this inclusive practice would not be enough. We must move beyond inclusion to the even more radical project of taking seriously in our own thinking the standpoint of the peoples of other cultures. The point of doing so is not only better to understand what other cultures' scientific and technological legacies have been and what they have meant to those cultures, but, of equal importance, to gain a more objective understanding of the North's achievements and what they have meant to those other cultures. We can see PCSTS as expanding the "sciences of sciences," as the Northern social studies of science and technology have understood their own work. The Northern studies can look at the integrity of moments in the history of Northern sciences with their particular eras (as Kuhn put it)—in this case, with eras of European expansion, currently carried out under the flags of modernization, development, and globalization.

This is to say that (with occasional exceptions) PCSTS theorists do not propose substituting a romanticized view of non-Northern scientific and technological traditions for the prevailing dismissive or demonized view of them which has accompanied the romanticized view of Northern traditions in Northern philosophy and science studies. Rather, they call for a more balanced, objective, "robustly reflexive" account of both. They want a critical assessment of the strengths and limitations of both kinds of traditions. They want accounts that take responsibility and accountability for knowable consequences of empirical research, but also for consequences which are difficult or impossible to predict, such as the effects of scientific and technological projects in one part of the world on the lives of peoples in other parts. Feminist postcolonial studies want women, too, centered in these projects as agents of knowledge and history, and women's lives considered as paradigmatically human as their brothers' in thinking about science and technology projects, a point to which we return in the next chapter.

Furthermore, note that this proposed new accounting system requires that non-Northerners also get to propose projects for global discussions. The North should (and will) have to negotiate the terms of scientific research, its applications, and how all of this is to be accurately and fairly understood. The North no longer is regarded as having the right to hold as uniquely legitimate its designs for possible future global science and technology scenarios—even ones intended to be progressive and inclusive, such as Latour's, Beck's, and Gibbons, Nowotny, and Scott's. Conceptual and political space must be extended so that the complex and conflicting perspectives of non-Western groups, too, can join in the public discussions of the history and sociology of science, and of science policy. Such processes must themselves be negotiated.

(2) New histories, sociologies, epistemologies, and philosophies of science

The postcolonial accounts raise new questions for histories, sociologies, epistemologies, and philosophies of science. How have scientific projects and traditions around the globe interacted with each other? What did each borrow? How has the West invented and maintained the notion of static, timeless, "traditional" societies? How have gender fears and desires shaped such projects? What do such histories do to exceptionalist and triumphalist philosophies of science? What happens to the cognitive core of Northern sciences if it has lost its unique legitimacy, as the PCSTS accounts argue? (Recollect the arguments of Nowotny, Scott, and Gibbons in *Re-Thinking Science.*) We saw Beck and Gibbons, Nowotny, and Scott raise this question from the perspective of what they characterized (respectively) as the ongoing demonopolization of sciences from the control of scientists and the valuable contextualization of scientific work. Yet neither analysis perceived the relevance of such arguments to the case of other cultures' knowledge systems. What can compelling philosophies of science look like when they fully engage with the PCSTS accounts?

(3) Multiple sciences: past, present, future

Importantly, such new accounts call for the recognition that there were in the past, are now, and always will and should be multiple scientific traditions which partially overlap and partially conflict with each other, as do their cultures more generally. Engaging more fully with this startling PCSTS insight is certainly one direction in which the new histories, sociologies, epistemologies, and philosophies of science must move. As indicated, for the North to ignore this is to promote a functional exceptionalism and triumphalism regardless of explicit attempts to distance such accounts from such positions.

(4) Relations between scientific and technological traditions?

But what then could be the future relations between modern Western sciences and technologies and the science and technology legacies and contemporary practices in other cultures? What should we teach our students about the sometimes conflicting claims that they make? (This presumes that we do teach our students about them!) Are there other and perhaps better questions to ask in such cases instead of only, "Which one is right?" Suppose we reorganized our science and technology courses as well as social studies of science and technology around "teaching the conflicts"? How could one do so while avoiding a corrosive epistemological relativism? We can start thinking about such issues by examining models of relations between the scientific and technological traditions of the North and the South which have already been proposed in the postcolonial accounts. Five such models can be identified.[17] Some of these will seem far-fetched and improbable when considered alone. I suggest that each should be regarded not as exclusively desirable, but rather as one element among the others in ideal future relations. But let us look at them in their original forms. We can keep in mind that it is extremely hard to imagine how scientific and technological work could be different than it is now. That is the power of the imperial legacy with its exceptionalist and triumphalist stances toward Northern scientific and technological traditions and practices. Thus, improbable and/or undesirable as one or another of these imagined relations between Northern and Southern science and technology traditions may seem, reflecting on the mechanics and possible consequences of each helps to envision possible futures different from those already familiar in the North.

Integrate other science and technology traditions into Northern legacies. One possibility is that useful elements of other societies' sciences and technologies would selectively continue to be integrated into Northern projects. The North has always "borrowed" from the cultures it has encountered. (Some think "appropriated" is the more accurate term.) This proposal recognizes the value of such integrations. And it recognizes that global forces beyond the control of any identifiable, sufficiently powerful counter-forces are in fact vastly increasing this practice. Efforts to accelerate this project are currently under way by Northern pharmaceutical firms eager to capture knowledge of cultures rapidly becoming extinct thanks to capitalist globalization processes. Maps for "mining civilizational knowledge" have been produced by one scholar from the South (and, no doubt, numerous corporations in the North) eager to see the achievements of Southern cultures better represented in a "global science" (Goonatilake, *Toward a Global Science*).

It certainly is important to preserve these unique and valuable contributions to the storehouse of human knowledge of nature and social relations. Shouldn't this and future generations have the benefits of both access to such knowledge and also awareness of its sources? Yet we can ask if this is the best way to insure such a future. Who should get to decide what should be preserved of other cultures' cognitive and practical legacies? Selectively integrating Southern scientific and technological elements would preserve only those aspects of other knowledge systems which can be incorporated into modern Northern sciences, destining to extinction other aspects attached to ontologies, epistemologies, ethics, and religious and other cultural commitments and practices which conflict with those of the North. Moreover, Northern sciences today are deeply implicated in corporate profiteering, militarism, and new forms of racism and imperialism, not to mention androcentrism, as we have already seen. And this happens in spite of at least some Northern scientists' explicit intentions to the contrary. Are these the people, scientists and science administrators, who should be making decisions about what to preserve from other cultures' traditions?

Moreover, every particular Southern cultural worldview left at the Northern laboratory door, so to speak, is precisely the origin of the insights and practices which the North, too, can now recognize as valuable once extracted from their cultural contexts. Should we be so confident that we know now exactly which parts of other knowledge systems are worth preserving and which should be permitted to become extinct? (Think, for example, how acupuncture was "integrated" into Northern neurology and medical practice only by leaving behind Asian conceptions of the body in balance and the health practices related to such conceptions.) Furthermore, this kind of one-way integration project would falsely seem to support unity of science assumptions and their ideal of one all-encompassing, legitimate knowledge system, exceptional and triumphant among the world's knowledge systems. This unity-of-science thesis has been thoroughly and widely criticized (cf. Galison and Stump, *Disunity of Science*). And what about the ethics and politics of this proposal? Is it acceptable for the North to let such cultures and, often, their peoples disappear while preserving only what the North finds useful for its own projects?

Delink. A second proposal argues that the sciences in the South should "delink" from Northern projects. In 1990, the Egyptian economist Samir Amin (*Delinking*) proposed that the economic systems of the South would never be able to flourish and to serve citizens of the South as long as those economies remained so firmly captured by Northern political agendas. Only complete withdrawal would permit Southern societies to develop on their own terms. The Third World Network put this kind of argument the following way: "Only when science and technology evolve from the ethos and cultural milieu of Third World societies will it become meaningful for our needs and requirements, and express our true creativity and genius. Third World science and technology can only evolve through a reliance on indigenous categories, idioms and traditions in all spheres of thought and action. . . . A major plank of any such strategy should be the delinking of the Third World from the secular dynamic which institutionalizes the hegemony of the West" (Third World Network, *Modern Science in Crisis*, 14). Certainly conventional histories of Western sciences and technologies routinely argue something similar about the ethos and culture of Western societies, though their exceptionalist and triumphalist assumptions are far more dangerous than those of a theoretically delinked Third World culture. The Westerners argue that modern (Western) science emerged in Europe because of the West's distinctive categories, idioms, and traditions, which originated in ancient Greek society, and were recovered for the European Renaissance. Why shouldn't Third World societies argue for their traditions in similar terms, albeit with a distinctive interpretation of their own exceptionalism and a bit of justifiable trimphalism, both distanced from the predatory and inperial politics within which Europeans made such claims?

Such a proposal seems unrealistic in today's world for two reasons. For one thing, it seems to be impossible now for a society to erect boundaries powerful enough to keep out transnational corporations, global media such as the Internet and cell phones, powerful transnational criminal activities such as arms and drug trades, and the expansion of terrorist projects. Since Amin proposed delinking, this project may have lost whatever plausibility it ever possessed as an actual political strategy. Second, pandemics such as SARS, AIDS, and Asian bird flu, as well as acid rain, desertification, and ozone holes, refuse to respect man-made borders. International cooperation is needed in attempts to head off such potential threats to human life and the life of the planet. So complete delinking, at least, seems both improbable and in important respects undesirable.

Nevertheless, it is certainly the case that the kinds of control that elites in the North exercise over the peoples of the South, intended or not, is not the only way, and in many respects not a desirable way, to conduct international relations about science and technology issues. What would a world of mutually, partially independent scientific and technological traditions and projects look like? Moreover, contemplating the effects of such a delinking brings into focus how dependent the purportedly autonomous North is on the "resource portfolio" provided by societies of the South (in Hobson's phrase: *Eastern Origins of Western Civilisation*). Where would the North's oil and mineral resources come from if not the Near East and Africa? Where would our cheap labor in Northern fields, factories, taxis, kitchens, gardens, and households come from if the South delinked from the North? Where would the bounty of supermarket fruits and vegetables and relatively inexpensive clothes and other manufactured goods come from if not from low-paid foreign workers? How could the travel industry survive without exotic Third World sites and workers, including its sex trade? Where would information technology workers, nurses, and laboratory scientists come from if they were not already birthed, brought up, and educated by other societies? Who would fill American university classrooms if all of the Third World students left? Clearly the "center" is at least as dependent on the "periphery" as the reverse situation commonly claimed by social theorists.

Integrate Northern problems, concepts, and practices into Southern science and technology traditions. A third proposal is that Southern scientific and technological traditions should flourish on their own terms while continuing the process of integrating elements of Northern scientific and technological traditions. After all, the globe's cultures have engaged in active trade with each other for at least 1,500 years, and this process has featured multi-direction borrowings, including scientific and technological concepts, practices, and projects. Northern technologies such as clocks and maps, medical practices, and agricultural and manufacturing techniques have also entered other cultures; the selective integration of one culture's achievements into others has long been a multidirectional process. It was the rise of the European empire that forced trade routes to shift toward Europe and, subsequently, North America, abandoning patterns of regional exchange which had nourished so many societies (Blaut, *Colonizer's Model of the World*).

Ambitious studies of these processes of integrating Northern sciences and technologies into non-Western societies have recently begun to appear. In one, historian Gyan Prakash analyzes the ambivalent relationship that India has had with Western science over more than a century (*Another Reason*). One focus is on the Indian projects to nationalize and Hinduize Western science by merging distinctively Indian and Hindu legacies, meanings, and practices with aspects of Western sciences. In another, Renato Ortiz ("From Incomplete Modernity to World Modernity") describes how in Latin America a distinctive narrative of modernization and its relation to tradition has given specifically Latin American meaning to such phenomena as urbanization, technology, science, and industrialization.

Notice that the combination of these first three patterns of relationship between Northern and Southern scientific and technological traditions[18] would create an interesting condition of rich global cognitive diversity in which societies sought mutual exchanges with each other while respecting each other's distinctive ways of living and coming to know natural and social worlds—at least those ways which did not harm their own societies. All societies could benefit from such exchanges. Cognitive diversity would be valued no less than biological diversity (Shiva, *Monocultures of the Mind*). We can at least begin to imagine the great value to all peoples of such conditions, even if they seem unattainable now.

Take other cultures as models for Western sciences and technologies. Fourth, Northern science and technology traditions could come to see other cultures as models for the North in significant ways. This would be a reversal of the conventional Eurocentric position that the North provides the uniquely admirable models of human achievement. After all, Southern societies have been having kinds of intracultural and intercultural experiences which the North has not, or at least has not recognized. They have overtly developed their knowledge systems not as culturally neutral, but as articulating some of their deepest ethical and spiritual commitments as, of course, Northern cultures did, however strongly they declaim the virtues of cultural neutrality. What if the North were to identify and root out elements of its knowledge systems which conflict with democratic ethical commitments and such values as a respect for nature and for what humans cannot know about its functioning? To take another kind of case, many Southern cultures have had the experience of everyday living in two conflicting cultures—their own and the culture of their imperial or colonial rulers. For example, educated classes in India learned both their indigenous health/medical systems and those favored by the British. They often learned to negotiate between the two. This in itself can be a resource.[19] However, its benefits may be far more extensive. Does learning to live in such ambivalence and ambiguity create valuable ways of thinking and engaging with others which are unavailable to those who think they can rely on their own natural superiority and their one right way to secure knowledge, health, a sustainable environment, and a rewarding life?

North/South collaboration. Finally, what about cooperation/collaboration between Northerners and Southerners in particular scientific and technological projects which respect the legacies of both Northern and Southern empirical knowledge traditions? Yet as long as Northerners have greater power, Southerners will be suspicious of such a relationship. As long as the Northerners disrespect the science and technology traditions of other cultures, such suspicion will be fully earned. Yet there no doubt are already and will continue to be contexts in which such projects can work. Identifying them is a promising route to valuable new ways of scientific and technological thinking and practice.

All but the first of these options may seem far-fetched to readers unaccustomed to thinking outside the exceptionalist and triumphalist narrative which is, we could say, the North's indigenous worldview. Yet I suggest that exploring these other practices and their possible consequences further can help us all to begin to imagine cultures of scientific and technological activity vastly different from the ones we have. We could begin to change our Northern tendency to permit our scientific rationality and technical expertise to serve inequitable power relations and to restrict what counts as legitimate knowledge claims to those which are fit for such service.

(5) Should science and technology studies become a site of public debate?

Let us first pose this question about the Northern studies. The histories, sociologies, and philosophies of Northern scientific and technological traditions have largely remained scholarly fields, struggling to find space within resistant departmental, university, and disciplinary professional organization priorities. They have created national and international associations and journals which enable the crossfertilization of ideas in different disciplines and on different topics. These new fields and the scholarly credentials of those who work in them are often devalued and resisted by powerful science departments which dislike and dismiss the critical, relatively objective gaze which these new fields cast on the traditional "folk-histories" of modern sciences and technologies. Moreover, Northern STS have favored lab and field studies that require support and cooperation from scientists. So it has become important to these researchers to keep open routes of cooperation with scientists.

The "Science Wars" of the late 1990s launched critical attacks from traditionalists, including Marxists, against the social constructivist tendencies in Northern science and technology studies. They did not focus on PCSTS, but they did have feminist work in their sights. This occurred at precisely the moment when the high Cold War funding for scientific research was sharply declining and neo-Liberalism was gaining power in the United States (Ross, *Science Wars*). These attacks stimulated extensive public discussion of science and technology studies issues. I say "public discussion," but the science warriors' charges were often presented more as hysterical performances than as thoughtful critical reflections. The science warriors did not understand social constructivist claims and construed them to deny that nature's order played any role in scientists' claims. That is, they interpreted science and technology studies as an extreme relativism. However, difficult and uncomfortable as these "debates" were for the scholars who were criticized, they enabled many progressive thinkers (inside and outside the sciences) who were unfamiliar with Northern science studies to come to a richer understanding of the constructivist arguments. I am not recommending more efforts by such groups to bring, in this case, postcolonial science studies to public attention through the kinds of hysterical public performances which characterized the Science Wars.[20] Yet because legacies of and current investments in imperialism, Eurocentrism, and androcentrism are central forces in the politics which shape science content and policies in the West, it is not enough simply to support a scholarly exploration of such issues. Western societies themselves need to engage in rethinking their identities as Westerners and as citizens of the world in today's multicultural and increasingly postcolonial and feminist world. Working through the PCSTS issues can make valuable contributions to such a project.

Western scholarly and research disciplines have long serviced the dominant groups through providing the conceptual categories and preferred causal relations between them through which public policy has functioned. They have provided the "conceptual practices of power" (see Smith, *Conceptual Practices of Power*). If that service to power is to cease, then it is not only scientific and technological decisions which must be returned to the political arenas where they belong, as Latour, Beck, and Gibbons, Nowotny, and Scott argued, but also discussions of the arguments in the "sciences of sciences" which provide new perspectives on research practices. Western scholarly and research disciplines must recognize the ways in which their conceptual frameworks become complicitous with political projects. They must seek public critical response to them precisely from those who would or could be affected by such frameworks. The modest retreat to disseminating such work merely in scholarly circles fails the accountability and responsibility test. They, too, must venture to negotiate with the feminist and postcolonial accounts rather than ignoring and devaluing them. This is a place where the courage of public intellectuals is needed.

To be sure, this is bound to be an unfamiliar and difficult process for science studies scholars. Those affected by the conceptual frameworks of such work are, on the one hand, the conventionally "excluded" from such processes, such as non-Northern cultures and the women and other marginalized groups in the North. On the other hand, another group affected is the already over-advantaged groups in the North and elsewhere which have benefited from imperial and Eurocentric scientific and technological practices. Whether or not science studies scholars intend their work to reach such eyes and ears, it will, as the recent Science Wars revealed. The response to unjust criticism should not be to purify science studies of its more radical thinkers, as some science and technology studies scholars have proposed. Rather, controversial positions have to be negotiated in public because they will affect public policy whether or not their authors so desire. Of course the terms of such a negotiation themselves must be negotiated; they cannot legitimately be imposed by only one side to any dispute. Polycentric democratic public discussion is needed. How can this be staged? Or, is it already being staged through some of the processes Beck as well as Gibbons, Nowotny, and Scott identify but seemingly restrict to the North?

(6) Modernity as a horizon for Northern science and technology studies?
Examination of the nature and implications of PCSTS throws into sharp focus how the "modern" in the Northern focus on modern science serves as a horizon for proper, approved, Northern perceptions and analyses. It prevents Northern gazes from straying over the boundary between modern and traditional to explore the nature of the traditional sciences of other cultures, their influences on Northern sciences, and the effects of Northern sciences on traditional empirical knowledge systems. Here we have looked at some of the central issues in postcolonial science and technology studies. However, there are additional issues which come into focus once one interrogates the modernity vs. tradition binary itself. Postcolonial studies have helped to prepare the way for this interrogation through support for the idea that there are and always have been multiple scientific and technological traditions; each culture in human history has developed a model that enabled it to survive (or not) the particular natural/social world in which it found itself. Yet the North's unique entitlement to an admirable model of modernization has rested squarely on the exceptionalist and triumphalist claims it makes about its sciences and technologies and their rationality—precisely the claims which PCSTS have undermined. We return to this topic in chapter 7.

6.

WOMEN ON MODERNITY'S HORIZONS

Feminist Postcolonial Science and Technology Studies

— — — —

WHAT ABOUT FEMINIST POSTCOLONIAL STANDPOINTS? Women in the South have distinctive standpoints on nature and social relations, including social relations with the North.

Men and women in every culture are assigned at least some different kinds of interactions with social and natural environments, although the gendered division of labor is sometimes far less gender-divided in pre-modern than in modern societies. But differences in gendered activities insures that imperialism and colonialism have had distinctive effects on women. The end of formal imperial and colonial rule seemed to offer possibilities for cultural self-definition and greater access to local resources. Yet half a century of so-called development projects have been controlled by Northern institutions and agencies, alongside global expansions of capitalist enterprises, both usually in league with elites in the former colonies. As we will see in later chapters, all of these agencies, institutions, and industries assumed that women's social needs and interests presented only obstacles to social progress. Under such conditions, equitable benefits have only rarely been delivered to women. These projects were intended to alleviate the impoverishment of the developing world. Yet women's impoverishment has increased during this period, both in absolute terms and in relation to men's conditions. Is this an outcome of defining as obstacles to "social progress" women's needs and interests? (Gender Working Group, *Missing Links;* Visvanathan et al., *Women, Gender and Development Reader*). Women have produced illuminating analyses of the obstacles they face and they have pursued innovative strategies for survival. So starting off thinking about knowledge production from the standpoints of Third World women's lives enables us to expand and, no doubt, revise our understanding of sciences "from below."

1. "THIRD WORLD WOMAN"?

In entering such a project, we must be alert to some problematic tendencies which we will encounter. Essentializing, homogenizing, and othering the "Third World Woman" are temptations to be avoided. As critics have pointed out, Northern feminist accounts have all too often posited an archetypal "Third World Woman" as the presumedly happy recipient of Northern feminist attention. Chandra Mohanty points out that such a figure homogenizes an immense diversity of women's situations, projects, and desires around the world. It reproduces the West vs. Rest structure of imperial and colonial thinking, leaving the "Third World Woman" the "opposite" of First World women. It positions non-Western women only as passive recipients of others' attention; it deflects attention from their own historical agency and present activism. It supports an uncritical attitude toward Western women's continuing service to racism, imperialism, and our own patriarchies. Western women are, once again, positioned as the "innocent," generous, and benevolent donors of attention on unfortunates abroad no less than at home (Mohanty, *Feminism Without Borders*. See also Narayan, *Dislocating Cultures;* Spivak, "Can the Subaltern Speak?").

Here we will focus on what can be understood about sciences and technologies, North and South, by starting off research about them from the lives of women from the South. Here we are setting out to reverse the characteristic "imperial gaze" and "male gaze" about which so much has been written. We intend to "gaze back" at Western imperialism and global male supremacy, especially as these appear in science and technology disciplinary and policy conceptual frameworks.

Yet such a project still retains the binary "Us vs. Them" which has had

such an inglorious history. The shifting and always Eurocentrically po-
liticized names which have been used recently for "us" and "them" and
which continue to be used here—developed/underdeveloped, First/
Third World, West/Rest, European/non-European, industrialized/
non-industrialized, and even "from above" and "from below"—signal
how problematic such a binary is. Many postcolonial scholars think it
inappropriate to continue to invoke these binaries which obscure how
the Third World is inside the First and vice versa. And they also obscure
many cooperative progressive projects between First and Third World
scholars, activists, social movements, and institutions which are under
way around the globe.[1] I agree that there are indeed important problems
with such binaries. Nevertheless, I think this "gazing back" is a valuable
project in spite of its always present limitations. Moving away from that
binary before fully appreciating its persisting legacy falls more under
the category of racist and imperial denial than of a progressive project
contributing to the dismantling of that legacy (Said, *Orientalism;* Willinsky, *Learning to Divide the World*). Preoccupation with the binary of
gender categories is equally problematic, while prematurely abandoning it also courts male-supremacist denial of its persistent efficacy.[2]

One must recognize also that there are men feminists and anti-
colonialists of European descent, as well as women anti-feminists and
people from former colonies who see themselves as having benefited
from colonial rule more than from the end of that form of governance.
Moreover, I do not assume that all women in the Third World are living
in pre-modern contexts. These issues are controversial, and must remain so in a world in which so many hope, but cannot feel confident,
that the future will be better than the past.

Of course imperialism takes different forms in different historical
contexts, and male supremacy varies not only in quantity from society
to society, but also in the ways it is institutionalized and enacted. What
we can do is remember that the reason there are and must be many
feminisms (as pointed out in the Introduction) is that women live
in different conditions, with different relations to patriarchal principles and practices, in different classes, races, ethnicities, and cultures.
Women around the world, like their brothers, have different and often
conflicting relations to both their own and the dominant sciences,
technologies, and modernities. Women's needs and desires often conflict with each other, and in the South no less than in the North. But,
then, who ever could expect men in different social circumstances to
achieve a unity in their perceived needs and desires?

One response to recognizing the diversity of women's conditions has been to restrict projects to publicizing "women's voices" from the Third World. These are valuable projects, but they could be stronger ones. It is still the Northern woman—or the woman from the South with access to such dissemination media—who is selecting which voices to present and which parts of their speech to reproduce. Moreover, sometimes such projects avoid the kind of hard theoretical and political analysis required to bring such speech to bear specifically as a standpoint on global social structures which are not easily visible from only local sites. Another strategy has been to engage only in analyses and projects relevant to some particular group of women—ethnographies, for example. This project gives up explicit universal claims for the solidity of particular ones. Such accounts can be illuminating and a valuable corrective to trafficking in false universals. Yet they also can be far too convenient a way to ignore the need to engage with the macro economic, political, social, and cultural forces which shape women's and men's daily lives. At the same time, such microanalyses can perfectly reproduce the global class, race, gender, and imperial relations which they were supposedly avoiding. Class, race, gender, imperialism, and cultural conflicts don't disappear if one ignores them.

There is no simple—or even complex—solution to these problems. There is no magical way to position an account such as this one so that it can be seen to be "right."[3] Nor is it preferable, therefore, either to retreat to the discredited positivist stance or simply give up trying to understand how others see us and themselves. Promising solutions can only emerge as inequalities disappear. Yet some encouraging tendencies already are visible.

The next section locates some central themes in feminist concerns within the framework of postcolonial science and technology studies more generally, which we examined in the last chapter. The last section identifies issues raised by this work for our project here.

2. THE NEED FOR POSTCOLONIAL FEMINIST SCIENCE AND TECHNOLOGY STUDIES

Only a few of the theories produced in the field of science and technology studies—postpositivist, feminist, and postcolonial—have treated Third World women as fully human. I am not saying that these theories make the outrageous claim that Third World Women are less than fully human—immature or deviant models of "rational man." Indeed, usually they do not mention women in the Third World at all. Rather their conceptual frameworks are shaped by androcentric and/or Eurocentric assumptions: they presume that men, whether of European or Third World descent, are the unique models of the fully human. Their periodization schemes, central analytic categories, and theories of social change, which shape their ontologies, epistemologies, and methodologies, block the possibility of any accounts at all of this vast majority of the world's women. Of course there are exceptions to these generalizations, as we will see shortly.

This is not an issue about the intentions of the scholars in this field, at least some of whom consistently support the interests of Third World women. For example, they actively and respectfully work to advance their Third World women graduate students and colleagues, and they do so in disciplinary and campus contexts which can be indifferent at best and even hostile to these women's interests and desires. In the Third World, science and technology intellectuals and activists have devoted immense energies to such projects as increasing literacy for women, including scientific literacy. Nor do I mean to belittle these important kinds of support or the courage and strategizing it takes to make them work. Rather, the issue is one of culture-wide elements of Eurocentric and androcentric conceptual frameworks which continue to shape work in this field. Projects in this field have not distanced themselves from the dominant frameworks in ways which permit actualization of the assumption that important insights about sciences and technologies of the First World and Third Worlds can be gleaned by starting off analysis from the standpoint of the daily lives of such women. None have taken these citizens of the world as models of "rational man," "manufacturer of knowledge," "revolutionary hero," "indigenous knower," preeminent exemplar of the "standpoint of women," or some other model of an ideal knowledge producer. Thus few science and technology studies accounts take the perspectives and issues of most importance to Third World women to be their own priorities.

This limitation is striking in the face of the long and often intense efforts of Western and non-Western national and international governments and agencies to shrink global poverty and to bring development to the "underdeveloped" Third World. Women and their dependents constitute the vast majority of the global poor. So one would expect to find that their conditions had improved as a result of all this attention and effort. However, during the last half century women's poverty around the globe has increased, both in absolute numbers and relative to that of men, as noted above. Clearly there is a gap between this increase in women's poverty and the widespread perception, at least in the West, that these agencies and organizations have been equitable in their distribution of development benefits.

Such perspectives and issues are the focus of concern in the work of many Third World women in both the North and South, as well as among others who have been working in national and international governmental agencies and in nongovernmental agencies on issues of how better to insure sustainable development for women in the South. A full survey of the literature here is not possible, but we can identify some of its themes which are valuable for this project.[4] (Those themes tied closely to issues of modernization theory and practice will also be addressed in chapter 8.) We will use the postcolonial science and technology studies framework of chapter 5 to organize these themes, and then direct attention to feminist issues within each category. That is, we will first try to "add women" to the postcolonial science and technology studies framework—an approach which we know in advance from earlier discussions will be both illuminating and frustrating.

3. ADDING WOMEN

Indigenous and traditional environmental knowledge

What kinds of skills, techniques, and knowledge do women need for their activities in "pre-modern" cultures?[5] First of all, in every culture women have some kind of primary responsibility for the household. This is so even when they are employed full-time outside the household, whether in the North or the South. Thus they have a responsibility for childcare and for the daily provisioning of household members in terms of their material needs, but also their social and psychic needs. They do this work themselves, or they are responsible for obtaining or managing the others who do this work, and the food, water, energy, and shelter such work requires. Such resources must be transformed so that they can be used in their particular households, and that means in the ways preferred in a particular culture.[6] Women also have responsibility for developing and maintaining their local environments—the kitchen gardens and other sites, such as forests, water supplies, and energy sources (e.g., firewood)—which make provisioning possible. They have identified, developed, and maintained the food sources necessary for survival (Kettel, "Key Paths"; Schiebinger and Swan, *Colonial Botany*).

But women's responsibilities do not end here. They are also responsible for the health, well-being, and daily and long-term survival of household members—children, the elderly, the sick, and the men who work in household contexts or leave the household daily to work elsewhere. They bear much of this kind of responsibility for kin networks which lie outside their households—in the North as well as in the South. Women typically are repositories of knowledge of their own and others' bodies, and how to make and keep them healthy in the presence of local threats to health and life. They have developed drugs, and medical and health techniques. In *Plants and Empire* Londa Schiebinger recounts the career of the Peacock flower in the Caribbean, which Indian and African slave women identified and developed as an abortifacient. The slave women routinely used it to prevent the births of children who would have to suffer the hideous conditions of slave plantations.[7] Schiebinger pursues the question of why knowledge of the abortifacient properties of this plant, widely known in the Caribbean, were *not* disseminated along with the plant throughout Europe in the eighteenth century.

In the context of these two kinds of responsibilities, women also have the kinds of moral decision-making and negotiation skills necessary to maintain social relations in the household. They do much if not all of the emotional labor. Moreover, women always play significant roles in community life. Their skills and knowledge for maintaining households, dependents, and social relations often are extended to maintaining communities. This is the case when women are not confined to the household but participate fully in mixed-sex community life. But it is also the case in sex-segregated communities where women have community lives of their own. Both mixed-sex and single-sex communities can be found in the North and the South.[8] This community work tends to be ignored when models of "the social" are restricted to "rational actors" and their actions, and to public, official, visible, and dramatic role-players and definitions of situations (Millman and Kanter, Introduction to *Another Voice*).

Furthermore, women around the globe work in many contexts be-sides households. They are craftspeople making jewelry, pottery, cloth-ing, baskets, blankets, and other artifacts. They are doctors and law-yers, and practice other professions which have indigenous elements insofar as they are defined and practiced within local cultural princi-ples, ideals, and legacies. They manufacture food and other products for use and for sale. They farm and do an immense range of agricul-tural work; raise and tend cattle, fish, and forests; and do whatever is necessary to insure the survival and flourishing of their communities, always within distinctive cultural contexts.

Such considerations show that women no less than their brothers need and develop distinctive kinds of skills, techniques, and knowl-edge for their daily lives in indigenous and non-Western societies. Yet most of the studies of indigenous and traditional environmental knowledge, illuminating as they are, take as exemplars the skills, tech-niques, and knowledge needed for men's activities in these contexts and societies. For example, they focus on navigational skills across the South Pacific (Goodenough, "Navigation"; Watson-Verran and Turn-bull, "Science and Other Indigenous Knowledge Traditions"; Hutch-ins, *Cognition in the Wild*), Native Canadian men's fishing and hunting (Bielawski, "Inuit Indigenous Knowledge"; C. Scott, "Science for the West"), and conventional notions of technological innovation and use (Adas, *Machines as the Measure of Man;* Headrick, *Tools of Empire*). Illuminating exceptions to such preoccupations can be found in stud-ies of Third World women's environmental, health, and economic practices (indicated in note 4), in mathematics (Ascher, "Figures on the Threshold"), in Japanese primatology (Haraway, *Primate Visions*), and in specifically gendered negotiations over technological change (Agarwal, "Gender and Environment Debate"; and the journal *Women, Technology, and Development*).[9]

Science and empires
The actual voyages, conquests, and initial institution of colonial gover-nance have been for the most part men's work. These practices are suffused with specifically masculine meanings (McClintock, *Imperial Leather;* Terrall, "Heroic Narratives"). Yet women too have played sig-nificant roles in these processes.

For one thing, European expansion, like Westward expansion in North America, left mirror-image sex ratio imbalances both in the settled worlds men left behind and in their new "pioneering" communities. In the European countries women had to learn to support them-selves in the absence of fathers, husbands, and future possible hus-bands. The men were out in the Spanish, Portuguese, British, French, or Belgian empires and colonies, often leading nasty, short, and brut-ish lives and leaving impoverished women and children behind. These women had to learn to navigate and negotiate in men's worlds, and they often developed institutions specifically aimed at relieving the impoverishment of women. What does this have to do with science and technology? Imperial adventures of conquest and discovery abroad were accompanied by different kinds of gender, race, and class strug-gles over power and knowledge at home. These two kinds of struggles have largely been located in isolated conceptual frameworks which obscure their linkages. Women, too, in their activities in Europe, were part of the voyages of discovery of Columbus, Napoleon's travels up the Nile, and Darwin's voyages on the Beagle.

But women in the lands Europeans encountered also played impor-tant roles in European expansion. Imperial and colonial governments often had policies about how the Europeans should relate to indige-nous women. Sometimes intimate relations and marriage were recom-mended in order to create ties between the foreigners and indigenous peoples. Indigenous women were to "suture" the foreigners into indig-enous cultures through such practices, producing bicultural children. At other times such relations were forbidden, to keep the foreign ad-ministrators uniquely loyal to the goals of imperialism and colonialism.

Several studies have documented the importance of women's family labor to indigenous efforts to resist the foreign imperialists and colo-nialists. The foreigners appropriated as much of the indigenes' labor as they could for their economic and political projects. But the labor they were least interested in or able to manage was that of reproducing, provisioning, and caring for indigenous children and other household members. Angela Davis (*Black Women's Role*) wrote about this in the context of slavery in the U.S. South. Mina Davis Caulfield developed such an argument about "cultures of resistance" to imperial and colo-nizing forces ("Imperialism, the Family, and Cultures of Resistance"). It takes diverse kinds of skills, techniques, and knowledge to live these lives no less than the lives of those who were at the time developing modern Western sciences and technologies to aid their imperial and colonial projects. (We return to the significance of this family labor for theories of modernization in chapter 8.)

Our topic here is women from the South. Yet we can also note that

European women in the South—the wives of foreign administrators, missionaries, and merchants, as well as those who were intrepid travelers—also played important roles in the imperial and colonial enterprises. Wives employed local women in their households, and in their dress and behavior were supposed to provide universally desirable models of womanhood to the local women. Furthermore, there were a good number of foreign single women, beginning in the eighteenth century, who went on their own projects and adventures abroad. Schiebinger recounts Maria Sibylla Merian's investigations in Surinam which resulted in hundreds of detailed engravings of native plants and of the insects in various life stages that lived on them. She was "one of the very few European women to travel on her own in this period in pursuit of science" (*Colonial Botany* 1). Some left travel narratives which reveal their relations to both the imperial and colonial projects and the indigenous peoples they encountered (Pratt, *Imperial Eyes*). Kavita Philip (*Civilising Natures*) recounts the continual mediation of British women and men in India in the "nature" they encountered in India through idealized images of the landscapes they had left behind in England.

In short, women engaged in many of the same kinds of knowledge production endeavors in colonial and imperial contexts as in their daily lives "at home"—whether this was Europe, Africa, the Caribbean, or India. Yet those contexts also provided different kinds of scientific and technological challenges to such endeavors. This work has only begun to be explored.

Women, the environment, and sustainable development

Aspects of these first two concerns—that is, knowledge production in colonial enterprises and in the indigenous cultures Europeans encountered—conjoin in feminist criticisms of the North's Third World development policies. By now, science and technology have been the critical focus of some four decades of struggle over what development could and should mean in women's lives, in contrast to how it has been conceptualized by powerful Northern-controlled national and international agencies. Since science and technology transfer from North to South has been a—perhaps even *the*—central concern of development policies for more than half a century, this has been a long struggle, first to understand how best to articulate such issues from the standpoint of women's lives and, second, to get national and international agencies to prioritize the findings of this work. This history has been told and retold in extensive detail. In retrospect, two issues have loomed especially large.

First, women were both left out of development, as the early analyses conceptualized the issue, and yet also, contradictorily, central to whatever successes development could claim, as later accounts pointed out (Braidotti et al., *Women, the Environment, and Sustainable Development;* Mies, *Patriarchy and Accumulation*). To be sure, women were from the beginning denied the financial support, education, and job opportunities their brothers received ("brothers" as representatives of "men" to acknowledge the class, race, ethnicity, and cultural inequities between men, too). This occurred as Northern scientific conceptions of agricultural, manufacturing, and infrastructural improvements were introduced in order to transform primarily subsistence economies into the profit-producing export ones desired by Northern-controlled transnational corporations and financial institutions. Development agencies used sexist Western conceptions of gender roles to portion out the opportunities and costs of their policies. Consequently, women were left behind in the development processes available to men in their societies as they were denied income-producing opportunities and at the same time assigned even greater responsibility for children, kin, and community care as their men were induced to work in modernization projects on plantations, in urban centers, or in mining, road and dam building, or other projects far from their homes.

Yet the situation was even worse. A major expansion of resources available in development processes came from the appropriation of the traditional land rights and labor of female and male peasants. This appropriation was by and to the benefit of the export economies which development processes worked to create. Processes similar to the "closing of the commons" several centuries earlier in Europe occurred in the early stages of Third World development, also. The farming, sylviculture, and grazing lands which supported peasant life became the private property of national and transnational corporations. Indeed such processes continue to the present day in, for example, Western corporations' appropriation of oil and mineral rights and pharmaceutical companies' appropriation of indigenous plant materials and of the indigenous knowledge which developed them into useful local pharmaceuticals. The former peasants, men and women, could now work for miserable wages for such employers, while the women also took on often sole responsibility for raising the next generation of workers. As even the international financial agencies have by now been forced to acknowledge, such processes further immiserated precisely the majority of the world's citizens whom development processes were supposed to help. They "developed" primarily the "investing classes" in the North and small elites in the South who were allied with the Northern corporations.

Such processes also advanced the destruction, sometimes irreparable, of the global environment upon which everyone's life and health depends now and in future generations. The kind of development we have had is self-destructive and cannot continue (Escobar, *Encountering Development;* Shiva, *Staying Alive, Monocultures of the Mind,* and *Stolen Harvest;* Seager, *Earth Follies* and "Rachel Carson Died of Breast Cancer").

Moreover, "nation-building" projects accompanying such development projects, in the South no less than in the North, often advanced men's benefits and rights as they restricted women's. Women lost prestige and status as they became second-class citizens of new states (C. V. Scott, *Gender and Development*). To be sure, many societies in the South already had entrenched, rigid systems of male-supremacist control of women. Yet development processes exacerbated such tendencies. (We return to these issues in the study of modernization policies and practices in chapter 8.)

A second theme, intertwined with the first, has been that development should not be defined only as increase in exports which profit primarily the already advantaged and destroy the material base of global life. It should include improving all aspects of human life— political, ethical, social, environmental, aesthetic, and spiritual (Braidotti, *Women, the Environment, and Sustainable Development;* Sparr, *Mortgaging Women's Lives;* Escobar, *Encountering Development*). Women were not alone in making this criticism, but their assigned responsibility for maintaining children, kin networks, and communities, coupled with the disregard that powerful national and international institutions had for their lives, gave them an especially valuable perspective from which to detect the limitations of dominant notions of development and to call for social change. Yet such a call could be effective only through persuasion. Women's groups around the globe have provided informational and conceptual resources useful to orchestrate their own demands and projects in international agencies, nongovernmental organizations, and national contexts. Today many are discouraged by the lack of response to women's demands for development policies that address their needs and interests rather than only the interests of Northern investors and their allies in the South. Yet these demands have radicalized communities around the globe. They have enabled and promoted the active agency of women and other disadvantaged peoples in analyzing and promoting scientific and technological change.

4. ISSUES: FROM "ADDING WOMEN" TO TAKING THE STANDPOINT OF WOMEN

Clearly, important additions to postcolonial science and technology analyses and activist projects can be made through the difficult work of trying to add women to them. Because men—as indigenes, travelers, invaders, revolutionaries, and historians—have not been interested in how women's activities have contributed to the kinds of enterprises they regard as fundamentally matters in "men's worlds," they have kept few records of women's activities. Consequently, just to find the traces of women's activities in order to add them to these existing histories is an amazonian task. Yet women were always present and engaged in their cultures' histories of finding strategies to insure the survival and flourishing of their children, kin, households, and communities. As we will see in chapter 8, they often have possessed more social status and power to bring to such projects in pre-modern societies than in the supposedly more progressive modernized ones. In such contexts women have struggled with changing physical environments and shifting social relations, and with encounters with other knowledge traditions. Systematic empirical trial and error, the forerunner of modern scientific methods, has always characterized attempts to understand nature's regularities, wherever they occur. Indigenous women's development and the maintenance of skills, techniques, and knowledge had to be dynamic, contrary to familiar Eurocentric and androcentric representations of such work as static, timeless, and mere repetition. Their thinking and practices were permeated by local cultural assumptions and beliefs, as are—and must be—the thinking and practices of the most modern of modern scientists today, as we saw in the first part of this book.

Moreover, colonialism and imperialism, in the colonizing nations and also in the colonized, created additional contexts in which women's scientific and technological ingenuity had to be exercised, whether in developing abortifacients and poisons (McClellan, *Colonialism and Science*) or integrating African and Caribbean botanical knowledge (Schiebinger, *Colonial Botany*). Some of the necessary additions to this work can be identified by using the gender lens developed for Northern feminist work to also examine women's lives in other parts of the world and to examine processes of North-South interaction, within which colonial and imperial practices must always have a central presence, though not necessarily always a defining one.

However, a number of additional issues are raised by this work. First, as long as real science and technology are identified only with the kinds of activities and concerns of interest to governments and corporations, from the design and management of which women have been excluded, it will be hard to see women as active agents in processes of scientific and technological change. Yet here and there they have been such historical actors. Thanks to the work of feminist historians such as Schiebinger (*Colonial Botany*), it is possible to see Maria Sibylla Merian as an important eighteenth-century scientist engaged in the same kinds of inquiries, and using the same kinds of methods, as her male colleagues. As Schiebinger makes clear, colonial botany was motivated, supported, and justified in terms of its service to national governments, and Merian's work also served such projects.

Moreover, in the second place, this framework used to identify "real science" turns out to be a Eurocentric one also, as we will explore further in chapter 8. Male supremacy can call on the resources of Eurocentrism, and vice versa, to defend against the full humanity of women and peoples not of European descent. It remains harder to see the agricultural, pharmacological, environmental, and medical knowledge developed by women in pre-modern societies as significantly scientific and technological because philosophies of modern science, even today, do not typically think of this kind of work as "real science" unless it is conducted in university laboratories by Western-trained scientists in white coats.[10] In her study of women in American science, historian Margaret Rossiter suggested that the professionalization of science in the nineteenth century was in part a direct response to—a reaction against—the increasing accreditation and achievements of women in these fields (Rossiter, *Women Scientists in America*). In

philosophy, too, professionalization has especially targeted the deval-
uation or exclusion of fields of study in which women's participation
has been high (O'Neill, "Invisible Ink"). We are entitled to wonder if
the persistent struggle by scientists and philosophers to maintain the
borders between "real science" and "folk sciences" is not in large part a
similar strategy of resistance to the increasing recognition of the empirical and theoretical achievements of other cultures' knowledge systems and to the high presence of women in the ranks of such knowledge producers, both at home and in other cultures. As long as the
household, women, kin and tribe, and "the local," as well as the "nature" associated with all of these, are conceptualized as obstacles to
social progress, male supremacy and Eurocentrism remain brothers
in arms against the rest of us. Incompetent and narrowly defined
and designed sciences provide important resources for their anti-
democratic practices.

A third issue in these writings is that development was supposed to
eliminate poverty in the Third World through the transfer of Northern
scientific and technological rationality to the lives of the "have-nots"
in the "underdeveloped" world. Yet the policy and practice of development has largely further impoverished the already economically and
politically disadvantaged of the Third World, of which women and
their dependents make up the vast majority. It has deteriorated their
environments and their communities, upon which their own scientific and technological legacies depend. At the same time it has excluded women from access to the benefits of the Northern scientific
and technological rationality available to their brothers. The consequent women's maldevelopment and de-development increases population growth and thus poverty in the South even further.[11] Nor is this
situation an accident. It appears that the "progress" required for the
success of the kind of development preoccupied only with increasing
export economies, which the First World had in mind, required the de-
development and maldevelopment of much of the Third World, including already poor women and their dependents. Feminists have
been in the lead in insisting that the concept of development be redefined to include social, political, psychic, ethical, and environmental
development, rather than only the narrow economic meanings and
references the term has had in Northern policy and practice.

Finally, women in many cultures in the South do not suffer as much
from the local reign of positivist ideals as do women in the North

(Narayan, "Project of a Feminist Epistemology" and *Dislocating Cultures*). Many live in cultures in which, whether or not the state is secular, religious belief is not regarded as antithetical to intellectual work. Thus the social and cultural character of knowledge production, in the "secular" North as well as the South, is more obvious to them (cf. Gole, "Snapshots of Islamic Modernities" and "Global Expectations").

This exercise in trying to add women to the postcolonial science and technology studies accounts has given glimpses of what the world might look like from the standpoint of Third World women. But to develop such an account further, we need to examine how the modernity/tradition binary has always been gendered, in the North and in the South. We turn first to recent arguments about the multiplicity of modernities, in the context of which we can better understand how modernity has always been masculinized and tradition feminized in the worlds which European modernity has created.

III.

INTERROGATING TRADITION

Challenges and Possibilities

7.

MULTIPLE MODERNITIES

Postcolonial Standpoints

— — — —

IN PART II THE SCIENCE STUDIES PROJECTS of groups on the periphery of Western modernity argued that there are and should be multiple scientific traditions and practices, each responding to needs and desires of its own local cultural and social context.[1] What implications could this have for conceptions of modernity? Could there be multiple effective scientific traditions and practices but only one which deserves to be called modern? Or is modernity, too, plural?

The science and modernity issues are internally linked for Westerners since what the West has meant by each requires the other. For Westerners, the term "science" is meant to refer only to modern Western institutions and practices. The term "modern" refers only to kinds of societies governed by the kind of rationality for which Western science provides the model. Consequently, as indicated in earlier chapters, the issue of multiple sciences either has only been hinted at or has been completely beyond the range of the intelligible in the work of many influential Northern science studies scholars. Of course significant contributions to thinking about this issue can be found in the work of Northern researchers and scholars.[2] Yet this pathbreaking work has not yet succeeded in shifting the conceptual frameworks of most postpositivist historians and sociologists of science, and certainly not of many philosophers of science. This chapter considers how the reality, desirability, and necessity of multiple sciences, argued in post-colonial science studies, become even more compelling when considered in the context of the discussions of multiple modernities. In turn, the multiple sciences arguments support the view that there are, and indeed must be, multiple modernities. Additionally, the modernity arguments enable a richer understanding of tradition, the pre-modern, and the social with which modern sciences are co-constituted.

Northerners have not yet fully confronted our deep commitments to triumphalist and exceptionalist understandings of the place in global history and social relations of our own forms of modernity, sciences, and technologies. Yet a pervasive illogicality suffuses such commitments. Critics point to the failures directly attributable to the North's modernity. And they ask how there could be only one modernity, or only one ideal form of modernity, given the multiplicity of ways in which societies transform their distinctive economic, political, and cultural legacies. Societies do so in response to three processes: changing natural environments, internal dynamics of their own social relations, and encounters with other societies' achievements and projects —voluntary and involuntary! So it cannot be only the West which can arrive at modernity under its own steam (as Westerners would have it). Moreover, every culture must always link the new to the familiar. So how could the modern ever completely replace the traditional? And just what is "the traditional" and the "pre-modern"? What are the intellectual and political commitments, intentional or not, that permit empirically and theoretically unsupportable notions of modernity and tradition still to function among citizens in the West who are overtly committed to reason and social justice? These are some of the questions central to recent studies of multiple modernities.

The next section reviews arguments made on behalf of such claims which we have already encountered in earlier chapters. Section 2 looks at some additional evidence in support of such claims. The concluding section identifies some puzzling and problematic issues raised in the multiple modernities work.

1. CONTESTING THE MODERNITY VS. TRADITION FRAMEWORK

The concept of the modern and the purported successes of the social practices it has directed have been vigorously criticized and debated in recent decades, some aspects of which I have reported in preceding chapters. Critics point to the horrifying failures of modern societies, such as their inability to address even the most basic needs for secure availability of food, health care, jobs, healthy environments, and freedom from violence for a large proportion of their own citizens. Meanwhile these industrialized societies export to the least advantaged peoples of the world even worse environmental destruction, poverty, ill health, and even more extensive violence. Defenders point to the great scientific, economic, social, and political achievements of modernized societies. They argue that it is not modernity's assumptions and practices that are responsible for the immiseration faced by people in both developed and developing societies. Rather, these are the consequence of the persistence of pre-modern conditions and practices, including superstitions, magic, and ethnocentric ways of thinking about nature and social relations. Sometimes they even seem to conceptualize the immiserated as little different from the "barbarians" and "savages" who populated Western perceptions of non-Westerners and the urban poor in their own countries not that long ago. It is only Western modernity which does and can provide the best hope for eliminating or at least ameliorating such obstacles to social progress, according to their claims.

Both images of modernity (if not the phantasms of barbarians and savages) can seem compelling. Yet simplistic understandings of modernity and modernization feed into both views and are disseminated in media and popular thought as well as through the global policies of transnational corporations and international organizations. Both the defenders and their critics share problematic assumptions about modernity and tradition. For example, as identified in earlier chapters, both claim that "modern" means Western and "modernization" means Westernization. Both assume that modernity thus can disseminate only West to East. And, equally problematically, both assume that West to East is the only direction of vision and insight through which one can rationally understand global social relations.

The defenders' vision of modernity was energized and directed by the enthusiasms of post–World War II social theorists and policymakers such as Alex Inkeles and David H. Smith (*Becoming Modern*), Daniel Lerner (*Passing of Traditional Society*), Talcott Parsons (*Evolution of Societies*), W. W. Rostow (*Stages of Economic Growth*), and Edward Shils (*Center and Periphery*), as well as by those of leaders of the American scientific community (Hollinger, *Science, Jews, and Secular Cul-*

ture). Recent reflection on the emergence of the social sciences in eighteenth-century and nineteenth-century Europe and on the formulations of modernity produced by Comte, Durkheim (Bellah, *Emile Durkheim on Modernity and Society*), Marx (Kamenka, *Portable Karl Marx*), and Max Weber (Runciman, *Max Weber*) provide one social crossroads about which late-twentieth-century social theory has had second thoughts.[3]

One insight stimulated by reflection on these debates is how necessary polemics about modernity have always been and must remain—how very modern they are! (Friedman, "Definitional Excursions"). Modernity has been contested and "in crisis" in the West from the moment it emerged. "The modern," "modernization," and "modernism" are used to mark actual, planned, or only imagined changes in direction for social relations and social institutions. Thus they also mark escape routes for some and threats to preferred social relations for others. As such, these terms are crucial sites where many groups can and do participate in debating the strengths and limitations of influential assumptions, institutions, intentions, and practices of the past, present, and future. Evaluations which appear compelling from one social vantage point invariably appear as merely rationalizations of local interests from other perspectives. For example, eighteenth-century and nineteenth-century debates about the virtues of European vs. North American models of modernization shaped discussions not only on those two continents but also on others, for example, throughout Latin America (Ortiz, "From Incomplete Modernity to World Modernity"). More recently, Third World societies have challenged First World conceptions of modernity, modernization, and their practices. Their responses are largely shaped by widespread perceptions of the failures of the First World's so-called development policies for the so-called "have-nots" and by perceptions of the falsity of exceptionalist and triumphalist Western self-images which always devalue non-Western traditions, including what we in the West can now begin to see as these other "traditions of modernity."[4] We have seen postcolonial critics focusing on the failures of scientific and technology "transfers" from the First World to the Third World and also on the discussions of the "oriental West" (Hobson, *Eastern Origins of Western Civilization*).

Then there is the matter of the impoverishment of the social sciences. As a number of observers have pointed out, we cannot expect them to guide us through today's concerns about modernity and modernization with critical perceptions attuned to the widespread disillusionments with both. The social sciences, their conceptual frameworks, and their methodologies arose from the same social processes that produced modernization in eighteenth-century and nineteenth-century Europe. They are themselves part of the modernization which is problematic even as they try to chart its nature and processes (Beck, *Risk Society*; *Reinvention of Politics*; Heilbron, *Rise of the Social Sciences*).[5]

Let us recall some of the main dimensions of the debates about modernity which are relevant to our project here. In the introduction, I distinguished histories of the conceptions, theories, and practices of three aspects of modernity: the modern in science and political philosophy, modernization in social theory, and modernism in literary and cultural studies.[6] Of course these discourses frequently become intertwined. Yet from their respective origins to the present moment they remain at least somewhat differently positioned with respect to the concerns of those particular disciplines. Nevertheless, in the heyday of each conception, one could see in those who embraced the modern the perception that the present or emerging moment was distinctively different from the past, that something new and valuable in social relations and social thought was emerging, and that the avant-gardes in each field were involved in that something new, even if they couldn't yet articulate just what its content was. Thus again and again, claims to modernity and its progressiveness were energized and became more plausible through a contrast with backward, intellectually and socially regressive tradition and the pre-modern. One could say that exceptionalist and triumphalist attitudes toward one's own culture seem irresistible to defenses of modernity.

The introduction also cleared some space for the discussions here by noting two distinct kinds of criteria used in defining the modern and modernization or their emergence. (Our focus is now on these two terms rather than also on the "modernism" of literature, the arts, and culture.) One is temporal, insisting that modern societies (or sectors of social relations) come after and replace traditional ones. But then one must specify some criteria for deciding the point at which the traditional becomes modern. Thus the temporal project seems to require a substantive specificity. And that has been a topic of controversy, since different observers have focused on different phenomena (Eisenstadt, *Multiple Modernities*). Political philosophers and social theorists often focus on aspects of modernity different from those that interest observers of modern sciences. Yet some common themes can be identified among the diverse substantive definitions of modernity. Modern societies are supposed to be forward-looking rather than focused on the past. Modern social institutions differentiate from each other. Thus the organization of the economy, governance, morality and religion, education and the production of knowledge, and the family come to have relatively autonomous principles and structures which differentiate them from each other. This kind of social organization differs from the kin-centered feudal and aristocratic societies in Europe's past and in non-Western societies. An instrumental rationality, epitomized by the rationality of scientific research, becomes the ideal in modern societies and comes to guide other institutions such as the law, education, and the economy. Yet plural conceptions of "the good" can be tolerated in modern societies and managed through democratic governmental elements such as a free press, multiple political parties, and popular elections.

Clearly, modern societies develop at different times in different contexts around the globe. Moreover, this uneven development is visible not only when comparing European with non-European societies. Parts of Europe were still feudal and aristocratically organized well into the twentieth century (Eisenstadt, *Multiple Modernities*). Official state religions still exist in a number of European countries, whether or not they are required or even actively engaged in by a majority of citizens. Already one can see that the contrast between modernity and tradition cannot be quite so neat and clean as the traditional theorists imagined. But the situation is even less clear than consideration of the temporal vs. substantive definitions of modernity suggests. We can focus on central problems with the contrast if we begin by looking at challenges to the temporal claim that the modern does and must always replace the traditional.

2. COULD MODERNITY REPLACE TRADITION?

In the multiple modernities literature, two lines of argument undermine the assumption that modernity does and could replace tradition, as the classical modernization theorists imagined. One looks at how modernity must always reproduce tradition as its Other. A second looks at how modernity must always be attached to the material, social, and cultural environments it enters, whether the modern emerges from within a culture's own historical dynamic or enters from outside the culture. The "suturing" of the unfamiliar to the familiar can be a useful way to conceptualize this second phenomenon. Thus modernities themselves do and must absorb traditional features of the cultures they enter. The modern is always changed by such interactions no less than is its pre-modern partner. Let us look at these two phenomena in turn.

The modern and its Other

Defenses of, calls to, and introductions of modernity must always reproduce conceptions of an undesirable past from which modernity offers an escape. Whatever else it may be, the modern marks a break with the past. So that past itself must be conceptually, symbolically, and materially carried along into the ideas and practices of modernity. Modern projects often grossly exaggerate the prevalence of pre-modern practices, in the process disseminating them far beyond their original borders. Moreover, what was traditional prior to the emergence of modernity is not necessarily what modernity refers to as traditional or pre-modern. Modernity invents and actively maintains material worlds and social relations which it attributes to the undesirable past and then regards as illegitimately persisting residues in the present (Giddens, *Consequences of Modernity*; "Living in a Post-Traditional Society"; Narayan, *Dislocating Cultures*; Philip, *Civilising Natures*; Prakash, *Another Reason*). We will return to this phenomenon when considering gender and modernity. Furthermore, it is not just the defenders of modernity who produce new forms of tradition. In the presence of modernity, those who resist it (one form of Beck's counter-modernities) also invent and actively maintain practices as traditional which often were less extensive, less intense, or even nonexistent in the past. This, too, will be a topic for the next chapter.

Thus "the modern" is always an oppositional term. To become attractive it must locate an "Other" worthy of its unfamiliar moral, political, social, and material demands. Yet at the same time as it defines itself against the (purportedly) familiar, and so depends upon the continuing presence of the traditional as Other, it also depends upon local traditions in other, more intimate ways.

Suturing

How can new things and ways enter any culture? In many ways. We can see some common features of such entrances by considering the example of technological change. Modernization is always perceived in terms of its technological advances, among others. Indeed, modernity theorists of both the left and the right have often seen technological change as driving modernization. This technological determinist view essentializes modern technology, identifying it with the history of Western industrialization. Yet social studies of technological change have emphasized the importance of at least four distinct aspects of such change processes which collectively show the impossibility of making essentialist and technological determinist arguments compelling. According to these social constructivist theories, technological change has four elements: (1) the introduction of new artifacts, along with the supporting material environments for their production, repair, and use; (2) new knowledge, skills, and techniques for designing, using, and repairing such artifacts; (3) changes in the social division of labor through which technological change can occur; and (4) new ethical, political, and social meanings of all of these changes.[7]

For most engineers and the general public today, the term "technology" refers to value-neutral artifacts. It has been easy for these groups to remain preoccupied with new kinds of such "hardware" to the exclusion of the other elements of technological change. Computers, spaceships, microwaves, electric automobiles, the OncoMouse, or Viagra seem simply to appear from engineering plans and the manufacturing processes which follow these plans. Such a preoccupation encourages a positivist and essentialist understanding of artifacts, in which their nature is set by their designers and the initial production process. Yet it has become clear that there is an "interpretive flexibility" to engineering products. Users frequently change their "natures" and meanings. For example, the telephone was initially intended for commercial purposes and the Internet for military uses. Yet both were turned into widely disseminated means for communication in daily life. Sometimes users develop technologies in directions directly opposed to the intent of their original designers. The elimination of paper was one of the goals of office use of e-mail. Yet electronic networking has not eliminated the quantity of paper copies which pile up in our offices. Instead, such networking most likely has increased the amount of paper on our desks, and this is not just because some of us old fogeys experience nostalgia for paper.[8]

Thus artifacts are not value-neutral; they can have politics (Winner, "Do Artefacts Have Politics?"). They do not have universal essences, and technological change consists of a great deal more novel social interaction than simply accepting the emergence of new hardware.

Neither the hardware nor its design and manufacture have sufficient powers to make change of artifacts an independent "motor of history" which itself can be held responsible for causing social change. Yet the design of artifacts does play a role in social change.

Technological change simultaneously requires a second project, namely, the development of new kinds of skills and knowledge—new "techniques"—for there must be people who know how to design, repair, and successfully use the new products. New kinds of schooling must be instituted and new ways for the new knowledge and skills to travel when formal schooling is inappropriate or insufficient. Often the new knowledge is not codified and so must be gained by bringing skilled personnel to new projects (as we saw Gibbons, Nowotny, and Scott argue in chapter 3).

Third, new technologies need welcoming physical environments. Telephones needed telephone poles, wires, sufficient electrical networks, household and business installations, switching offices, telephone books, and telephone bills. Construction of these environments required its own skills and knowledge, and training programs to develop them. Cars required roads, gas, service stations, road maps, signage, new lighting systems, parking areas, and eventually traffic cops, parking meters, driving schools, state licensing agencies, international road symbols, and pollution control laws. These days we can watch the debates over how the various alternatives to fossil fuels currently being developed—wind farms, solar systems, ethanol, electricity, and fuel cells—are to be sutured into their physical environments, as well as into cultures in other ways.

A fourth aspect of technological change always concerns the struggles over which social groups will get to design, use, or repair new technologies and their required material environments, and which ones will benefit from these. Who will get the highly skilled, highly paid, and high-status educations and jobs and who will end up with the minimally skilled, low-paid, and low-status jobs, or even with no jobs? Which groups will benefit and which will lose as typists are replaced by data entry technicians, as women and non-U.S. citizens seek access to jobs as astronauts, as manufacturers try to figure out whether microwave ovens will become another kitchen machine integrated into the cook's food preparations or merely a way to heat tea or provide minimal resources in bachelor apartments, and as new forms of scientific research move from universities to industry?[9]

Finally, in order for any of these changes to take hold, potential users must become excited about—or at least not hostile to—the meanings of a particular technological change. Of course new technologies have an instrumental rationality; for example, they can enable better business or military communication. Yet they also have "interpretive flexibility" such that their expressive meanings are context-dependent. Processes of technological change are enabled and limited by the ways that an artifact and its uses can be linked to local values, interests, and practices. People have to come to want—to value—the new artifacts and the kinds of knowledge and social relations enabled by their ongoing design, use, and repair. They have to see the new artifact as performing a social function they desire. They have to make it meaningful for their daily economic, political, psychic, and spiritual life. For example, as indicated above, the telephone and Internet became desirable to women, especially, as ways of maintaining social and kin networks in everyday life.

The Japanese philosopher and historian of technology Junichi Murata ("Creativity of Technology") identifies this process with respect to the locomotive and rail system introduced in Japan in the late nineteenth century. "Although modern transplanted machines such as [the] steam locomotive and railway system did not function successfully in the sense of instrumental rationality, they had a great expressive meaning as a demonstration of Western civilization in the early Meiji era. . . . A train pulled by steam locomotives could be considered a kind of running show window or advertisement media for modern Western civilization. . . . [People were] motivated for a certain interpretative activity and began to 'see' the modern Western world 'through' a train" (258; cf. Prakash, *Another Rationality*).

Murata brings together these various ways of suturing modernity to tradition in his discussion of the importance of the traditional technology sector in modern Japan. The creativity of an advanced sector of modern technology has depended upon the restriction to culturally acceptable patterns of the flexible ways a technology may be interpreted or linked to local values and practices. This adaptation is provided by the traditional technology sector in Japan. "One of the most conspicuous characteristics of the modernization process in Japan is the dual structure of its sociotechnical network with an advanced sector of modern technology and a parallel domestic sector of traditional technology. The advanced sector functions as if transferred technology guides and determines the way of modernization. In reality, however, the advanced sector interacts with the domestic sector, where traditional technology plays a role of instrumental rationality, decreasing the gap between the two sectors sufficiently that advanced technology becomes adapted to local practices. Through this interaction, the scope of flexivity is restricted, the process is channeled in a certain direction, and rapid and continuous adaptation and development of technology becomes possible" ("Creativity of Technology" 263). Counter to the claim that modern technologies contrast with traditional ones in that the former are context-free and therefore universal, the Japan case makes clear that "without an environment provided by traditional technologies, modern technologies cannot be transferred and introduced into other contexts. In this sense, we could say it is rather the developmental processes, translation and transformation processes of traditional technology, that make the modernity of technology possible. Without such support from traditional technologies, the ideological character of modern technology could not be transformed into reality. Modernity, without the help of tradition, would remain only an ideology" (262–63). Thus "in modernity there is always a dual structure constituting modern factors and traditional factors. In this sense there are always various modernities (plural) together with various transformational processes of tradition" (263).

It is important to understand that these suturing processes, in which a new technology is enabled to function in a particular context, are changing "the modern" and even constituting the technological change itself. Thus the modern absorbs or is shaped by the environments it enters; it does not stand apart from them attached by tubes and stitches which leave intact an original modern feature. Perhaps this issue seems obvious in the preceding discussion of suturing new technologies into their new environments. But sometimes the suturing process is conceptualized as simply changing everything in the environment of a new technology but leaving the artifact—that piece of modernity—itself untouched. That is an inaccurate way to understand such processes.

3. SCIENCE IS TECHNOLOGICAL

I have focused on technological change here because every theory—left, right, and center—takes it to be a central feature of modernization. Other cultures are recognized as having technological practices, but not scientific ones. That is, the boundary between pure science and its applications and technologies is implicitly preserved even in otherwise progressive accounts of knowledge systems around the world. The retention of this boundary is another artifact of modernity—one of the "great binaries" of modernity, as Nowotny et al. put it in *Re-Thinking Science*—and it is not useful for our project here. The social constructivist approaches to the history and sociology of sciences and their technologies show these two human activities to be much more tightly linked than the conventional philosophies of science and technology could ever imagine, as the theorists considered in Part I showed (cf. Biagioli, *Science Studies Reader*). To say this is not to imagine that it is always useful to think of scientific and technological change as identical processes or ones with completely parallel histories. Yet once the concept of technology is no longer restricted to hardware but includes the three other aspects of such change, and when science is taken to consist not just of abstract representations of nature but also as distinctive kinds of interactions with it, then contexts begin to appear for examining how they function together as one. Science studies has convincingly demonstrated that the purported boundary between them, if it ever existed, has now been largely dissolved.

What kinds of dissolutions of that boundary have been identified? First, scientific questions can have practical origins: Is there a remedy that will cure cancer or AIDS? That will reverse global warming trends? Such origins of research problems are not thought to damage the autonomy from social elements of the methods or content of scientific claims. According to conventional beliefs, there are lots of questions one can ask about the world around us; the fact that some of them seem urgent or just scientifically interesting is no reason to think that the social neutrality of the results of scientific research will be breached. Yet cancer, AIDS, and global warming "have politics," as we can observe from the struggles of environmental and medical researchers over funding for cancer research; the struggles of pharmaceutical companies, gay health groups, women's health groups, and poor people's movements over AIDS research; and industry, military, and governmental resistance to proposals to reduce global warming.

A second relation between science and technology is that the results of "basic" scientific research can be used to create practical applications and technologies. This is the link on which defenders of the autonomy of science have usually focused, arguing for the social neutrality of "basic research." Indeed, much of the research on cancer, AIDS, or global warming must still be basic research, defenders of the autonomy thesis argue. We do not know enough yet about these phenomena to apply such knowledge in particular contexts or to construct technologies of cure or prevention. Yet we saw both Beck and Gibbons, Nowotny, and Scott focus in different ways on how the patterns of funding and sponsoring scientific research today make it difficult to identify any socially autonomous research at all. "Basic research" has to be paid for by someone. Which "basic" questions, of all the possible ones in the universe, such research should address and how best to address them are not themselves autonomous from the social interests and desires of funders and sponsors. Poor people's health and environmental issues tend to get little attention (unless such issues involve the ability to work or the possibilities of contagion); rich people's issues get a lot of attention. Of course, individual scientists may well pursue their research blissfully ignorant of the uses to which their funders and sponsors intend to put it. But the ignorance of scientists doesn't make their work autonomous from the interests and desires of their funders and sponsors.

A third kind of dissolution of the boundary between science and technology has been recognized in the fact that scientific research uses technologies—including hardware, techniques, organizations of research labor, and meanings of methodological elements—to collect and evaluate data. It uses astrolabes, telescopes, microscopes, maps, sailing ships such as Darwin's, the earlier ships that enabled "voyages of discovery," and many more devices. However, which technologies a culture can and is willing to use is a matter of social and cultural history. The design, use, and repair of each requires new skills and techniques ("methods"), new supportive material environments for such processes, new ways of organizing the labor of scientific research, and attractive new meanings of such processes (cf. Barad, "Getting Real" and *Meeting the Universe Halfway*; Latour, *Pasteurization of France*). As long as technological change was conceptualized only as the construction of new and culture-free artifacts, the autonomy of science from the research technologies it encountered could seem socially and politically innocent. But once constructivist accounts of technological change appeared, the picture changed. Now scientific methods themselves are reasonably conceptualized as technologies of research. Distinctive methods of collecting and evaluating empirical evidence are supposed to be defining aspects of scientific change, so this is no small change in conceptualizing links between technologies and sciences or, rather, rethinking just what is meant by "scientific methods." Scientific research is technological, and therefore socially constituted, at its cognitive, technical core. (This is not to say that nature plays no role in scientific practices, as some critics of constructivism have assumed, but only that culture, too, shapes the results of research.)

Latour's study of Pasteur's research provides an influential example of this kind of account. Pasteur's project required new research tools, new kinds of knowledge and skills, new physical and social environments for the research, the recruitment of farmers and physicians as participants in scientific research, a reorganization of the labor of health care workers, and the dissemination of culturally acceptable meanings for these changes. Scientific innovation depended upon carefully orchestrated technological change, he shows. Or, rather, scientific change and technological change were co-constituted and came into existence simultaneously with broad changes in French social relations. In chapter 3 we saw Gibbons, Nowotny, and Scott propose that the forms of mission-directed research occurring in industry were both "manufacturing" products and also contributing to the growth of scientific knowledge. A particular piece of research could not be categorized as only technological or only scientific. A central theme in their complex narrative can be understood as the account of how successful new forms of research "suture" science projects into their material, social, and cultural environments in ways that make it impossible to isolate technological from scientific change.

The interactionist tendencies in philosophies of science increase the relevance of appreciating the value of thinking of scientific work as technological at its core. Conventional epistemologies and philosophies of science have tended to conceptualize science as fundamentally a set of representations of reality. This representational view, like the conventional view of technology as merely hardware, insures the isolation of these accounts of science from their social and technological contexts, but at the expense of our ability to understand how scientific and technological change occur. Thinking of the goal of scientific work as, among other things, the successful interaction of scientists and their technologies with material, social, and cultural contexts enables philosophies of science to make use of more of the resources created by social histories and social constructionist sociologies of scientific and technological changes.

Thus we can say that sciences, too, have a kind of interpretive flexibility. The very abstractness and generality of modernity's sciences and technologies permit them to be interpreted in many different ways and thus to be practiced, applied, or interpreted in many different cultural contexts. But in each case, such abstract and general principles must be integrated into—sutured to—local physical and social environments, and cultural resources, values, and interests. This task can only be done through traditional craft labor. And the suturing changes everything involved in the process, whether it entered as part of tradition or part of the modern.

I have been developing the arguments for multiple modernities through examining the processes and effects of suturing modern sciences and technologies into "traditional" contexts. As a result of such processes, modernity will be different as it is adjusted to fit into each context it enters, and so will its sciences and technologies. I have been identifying a kind of logic of the emergence of modern features in any particular context, drawing on the rich case studies to be found both in science and technology studies and in the multiple modernities studies. This work raises several additional intriguing issues.

4. ISSUES

Here I will focus on just three issues: the sources of multiple modernities, the issue of disappearing modernity, and the possibilities for interrogating "tradition."

Multiple modernities: two sources

The modernity theorists believed that the substantive features of modernity could, would, and should be disseminated from the West to other societies around the globe. Only in the struggle toward the modernity modeled by the West could global social progress be measured. Thus, slowly, global culture would become homogenized as modern features replaced the plethora of particularistic features found in traditional societies. Yet, as we have seen, homogenization has by no means followed the dissemination of Western modernity. Consider its sciences. Instead of "international science" replacing traditional knowledge systems, we see instead a world of indigenous, contextualized knowledge systems, in which modern Western science is simply one among many, albeit an extremely influential and empirically powerful one. It turns out that the fact that a particular scientific tradition can travel "internationally" cannot be used to support the view that it has no distinctive cultural or historical features. To the contrary, the arguments coming from science and technology studies show that it brings distinctive features of the originating culture into the recipient culture and also must take on features of that recipient culture as a requirement for actually becoming operational there.[10] Indeed, valuable elements of the scientific traditions of other cultures have themselves disseminated to the West, enabling the advance of Western sciences (Goonatilake, "Voyages of Discovery"; Hobson, *Eastern Origins of Western Civilisation*; Weatherford, *Indian Givers*). So this kind of cultural borrowing, in every direction, is one of the sources of plural modernities.

But the heterogeneity of modernities has another source. This is the independent emergence of modern features within other societies, both before and during the advent of European modernities. Everywhere ancient physical, social, and symbolic materials are recycled; everywhere environments undergo changes and societies must learn to survive under new conditions. Here the work on early modernities around the globe is especially useful (Eisenstadt and Schluchter, *Early Modernities*) as is the complementary work on Eastern origins of Western civilization (cited above). In these accounts, European modernity was created through a combination of internal processes and extensive borrowing from societies more advanced than Europe by the early fifteenth century. The internal and regional dynamics of other societies produced rich and complex modernities of their own, some elements of which are responsible for the rise of European modernity. The Eurocentrism endemic in the West has obscured these histories even from the sight of progressive scholars in the West. Moreover, the economic, military, and cultural power of the West in the last century or so has further entrenched "the winner's story" of the history of modernity. In contrast, in these new accounts, modernity was multiple already before Europe arrived on the stage of world history, and the emergence of a distinctive tradition of modernity in Europe has not changed that history. What will histories and philosophies of modernity and its sciences, including their epistemologies, look like when they recognize and integrate these counter-histories?

Is modernity disappearing?
Modern societies are supposed to require and in turn constitute differentiated social institutions, according to the standard account. The

organization of the economy, governance, morality and religion, families, education, and science are supposed to have principles and structures which are independent of each other. Yet we have seen several of the science studies scholars wondering if modernity is disappearing. They note that science has appropriated political decisions which belong in public-sphere democratic processes while also infiltrating public policy discussions. They mark how political goals increasingly are shaping scientific research. They identify how industry economic goals and standards are shaping scientific research, the research located in both industrial settings and universities. And there are more such incursions of one institution into another which appear in this work. Attention to the "contextualization" of scientific research, in Nowotny et al.'s language, highlights how Western institutions today are both permeable by other institutions, and in turn transgressive with respect to them. Of course, critics have pointed out that both Beck and Nowotny et al. overemphasize the break in the history of modernity which has recently occurred. So was this original institutional disaggregation claim at least in part a piece of public relations advertisement for modernization in the face of its actual failure completely to disaggregate its institutions? Perhaps it was, in the same way that the autonomy of science was proclaimed in the United States in the 1950s precisely when the federal government was becoming much more deeply involved in supporting and regulating scientific research (Hollinger, *Science, Jews, and Secular Culture*).

Is modernity disappearing as its major social institutions de-differentiate or re-aggregate? Or, is it only the dominant Eurocentric and androcentric view of modernity which is declining as we get insights into what kinds of institutions could serve the vast majority of the world's citizens who were always excluded from the category of fully human, ideal subjects of modernity? We return to this issue in the next chapter.

Interrogating tradition
Finally, the discussions of multiple modernities bring into focus the necessity of also problematizing the category of tradition. This project was unintelligible as long as modernization was identified as Westernization (Giddens, "Living in a Post-Traditional Society"). We noted above that the tradition which existed before modernity appeared in a particular social context is not necessarily identical to either the tradition against which modernities define their appeal or the tradition

which they must reproduce in order to function in any particular context. Modernity produces "tradition" in several ways. Invoked as modernity's Other, tradition calls forth deep anxieties in Westerners, especially elite Western men, about their conceptions of modernity and modernization. As we will see in the next chapter, to separate from tradition is to leave behind women, femininity, the household, loyalty to kin and tribe, as well as nature and "the primitive." Under such conditions, could women ever be modern? Must not the gender of modernity and tradition be confronted directly if this binary is to be successfully abandoned?

The kinds of rational, critical counter-histories recounted above and in earlier chapters do not yet engage with the emotional charge of the modernity vs. tradition binary in the North. What are Northerners so committed to denying through insistence upon this contrast? Why do defenders of traditions seem to experience threats to the integrity of their traditions as threats to their deepest senses of self? Why do commitments to tradition as well as to modernity seem so compulsive?

8.

HAUNTED MODERNITIES,

GENDERED TRADITIONS

— — — —

NO LONGER CAN MODERNIZATION be thought identical to Western-ization. No longer can one reasonably think there has been, is, and can be only one model or set of practices of the modern. No longer can modernity be assumed to replace tradition or the pre-modern, suc-cessfully banishing these to the past. No longer can the very contrast between the modern and the traditional be thought to reveal more than it hides. We have arrived at these conclusions by thinking about mo-dernity and its sciences from the standpoint of non-Western cultures and from the multiplicity of histories that modernity has had within Europe and North America.

What about gender? Women have been noticeably absent from or marginalized in the paradigmatic sites of modernity—the politics of the public sphere, science and technology, direction of capitalist enter-prises, and leadership in civil society. Is this purely an accident? Mod-ernization theorists treat it as a residue of traditions that moderniza-tion is eliminating. The West's women are claimed to be the most modern, and thus the appropriate model for women around the globe. Western feminists have often made this assumption also. But is wom-en's exclusion from the paradigmatic sites of modernity merely a residue of traditional gender relations? Or is the modernity/tradition contrast far more deeply gendered? Can the social sciences help us out here in critically examining if and how this contrast is gendered?

The social sciences were formed through studies of the nature and social effects of urbanization and industrialization in Europe and North America which the founders of social science saw all around them. That is, the social sciences were themselves part of that very era of modernization which was the specific topic of the great nineteenth-century classical modernization theorists (Heilbron et al., *Rise of the Social Sciences*). Whether explicitly or implicitly, the contrast between modernity and pre-modernity, as these theorists experienced it, came to shape the founding conceptual frameworks of the social sciences. And gender relations were not even perceived as social matters by these theorists or in the social climate of their day more generally. This insight helps to explain two things. First, one should not expect mainstream social sciences to be of much help today in critically moving beyond the modernity vs. tradition framework, as Beck, for example, pointed out. Second, we should not expect much help from today's mainstream social sciences in identifying gendered aspects of modernity to which the founding modernization theorists were themselves almost completely blind. Well into the last half of the twentieth century, the conceptual frameworks of mainstream social sciences made gender relations external to the social relations they studied, as more than three decades of feminist social sciences have shown. Male supremacy has not been treated as a social matter, let alone as an undesirable one. The social sciences have treated male supremacy as either biologically determined or as a trivial matter of individuals' behaviors. These research disciplines have lacked the conceptual resources (not to mention the will!) to explore how central gender relations have been to historically specific social relations in different societies and in different eras. Women rarely make appearances in social science accounts of social progress. Even more rarely are ideals of social progress critically evaluated from the standpoint of women's lives.

Northern feminist philosophies of science have pursued two critical directions in reevaluating the movement from tradition to modernity in which modern sciences and technologies have played such a central role. Alongside other feminist science studies, they have pointed out that women have made important contributions to what is thought of as pre-modern empirical knowledge of nature and social relations. Yet this knowledge is often more reliable than that certified by mainstream scientific institutions, such as medical and health institutions and the social sciences. Additionally, they have identified how those faulty modern ideals of scientific rationality, objectivity, and good method are shaped by a familiar stereotype of manliness, as we saw in chapter 4. Yet there is a third strategy worth pursuing. This one begins from women's experiences of Western modernity and interrogates the way gender stereotypes constitute an important dimension of the very project of the contrast between tradition and modernity. The value of such a project is implied but not pursued in Northern feminist philosophies of science.

Here I focus on the emergence of this third kind of critical perspective on modernity's ideals and practices. The final chapter proposes one methodological strategy which can enable powerful alternatives to the West's male-supremacist and Eurocentric notions of modernity, tradition, and social progress. Here we begin, first, with an early feminist argument about the radical effects that recognizing women's equal humanity can have on the conceptual framework of historical thought. Thinking about modernity is by its very nature thinking about the passage of time. We will then pursue how to use these insights about history to start from women's experiences of modernity to reevaluate the modernity vs. tradition contrast.

1. METHODOLOGICAL SUSPICIONS

Joan Kelly-Gadol pointed out more than three decades ago the radical implications for history of assuming that women, too, are paradigmatic of the human; that, in her words, "women are a part of humanity in the fullest sense" (Kelly-Gadol, "Social Relations of the Sexes" [SRS] 810).[1] This seems like an absurdly obvious belief. Yet it is far too rarely exhibited even today, after more than three decades of feminist biology and social science research. Kelly-Gadol argued that when one makes the assumption that women are fully human, standard features of historical analysis, such as periodization schemes, categories of social analysis, and theories of social change, no longer look reasonable. From the perspective of these arguments, we can ask what happens to scholarly and popular assumptions about modernity and tradition when one regards women, too, as fully human.

Periodization schemes

Kelly-Gadol has pointed out that feminist historians

> [had already begun] to look at ages or movements of great social change
> in terms of their liberation or repression of woman's potential, their
> import for the advancement of her humanity as well as "his." The mo-
> ment this is done—the moment one assumes that women are a part of
> humanity in the fullest sense—the period or set of events with which we
> deal takes on a wholly different character or meaning from the normally
> accepted one. Indeed, what emerges is a fairly regular pattern of relative
> loss of status for women precisely in those periods of so-called progres-
> sive change. . . . For women, "progress" in Athens meant concubinage
> and confinement of citizen wives in the gynecaeum. In Renaissance
> Europe it meant domestication of the bourgeois wife and escalation of
> witchcraft persecution, which crossed class lines. And the [French] Revo-
> lution expressly excluded women from its liberty, equality, and "frater-
> nity." Suddenly we see these ages with a new, double vision—and each
> eye sees a different picture. (SRS 810, 811)

Conventional histories have claimed, without evidence, that women
shared with men the benefits of such progressive advances. Yet—Kelly-
Gadol is arguing that in fact women's lot usually has worsened in eras
regarded as progressive.

Moreover, the situation is even more discouraging. The restrictions
on women which occur at such moments of supposed advances for
humankind are not accidental, but rather "a consequence of the very
developments for which the age is noted" (SRS 811). As Euripides
recognized in *Antigone,* women's social power, which in traditional
societies is exercised through responsibility for kin and community,
must be weakened and even destroyed for democracy to emerge. In
ancient Greece, it was replaced by the power of the city-state—Athens,
in this case, protected by the goddess of wisdom, Athena. We are told
that she was "not of woman born" but rather sprang full-grown from
the brow of her father, Zeus—an evidently significant detail that has
frequently caught the attention of feminist critics of Liberal democratic
theory and its conception of legitimate knowledge. Already we can see
here the theme of progress toward democracy, which would some two
millennia later become a significant mark of modernity, defined in
terms of masculine escape from and struggle against the responsibilities of family relations and the powers of women.

The political philosopher Carole Pateman focused on the internal links between the founding of democratic states in Europe and North America and women's loss of power and prestige in these processes. These links are articulated in Liberal political and legal philosophy: for the privileged classes of men, only their greater control over women can make up for having to accept lower-class men as political equals (Pateman, *Sexual Contract*). For Pateman, the purportedly progressive social contract is also a sexual contract between men and against women. Anthropologists, political scientists, and historians who examine the processes of state formation more generally consistently note how women consequently lose power and status, as they also do in the emergence of democratic states from aristocratic feudal societies, which has invariably been hailed as a progressive moment. Nation building has been a widespread postcolonial political project in the Third World during the last half century, and neither modernization theorists nor their severest critics, dependency theorists, have effectively grasped women's situation in such projects (cf. C. V. Scott, *Gender and Development* [GD]).[2] We return to this point shortly.

Every social transformation redistributes material and social resources—the benefits and costs of historical change. So calling any historical era progressive without regard to the situation of women or of other already economically and politically disadvantaged groups seems a practice best understood as "the winner names the age." It is a questionable practice on moral and political grounds when the disadvantaged groups constitute only a minority of a society's (or the world's) peoples. Yet when women and their dependents are at issue—and those who depend on women for their survival or care suffer when women lose power and prestige—the disadvantaged will always constitute a vast majority. Without direct attention to the full humanity of the worst-off, whatever their numbers, putatively democratic struggles cannot escape providing their benefits only for the few. Full attention to the consequences of considering the worst-off to be just as paradigmatically human as the best-off can no longer be considered an optional project when assessing which historical moments and processes are progressive "for humanity." Yet when getting powerful men to accept equality with less powerful men requires a payback of more power over women, clearly democratic projects are facing something worse than mere oversights or residues of traditional forms of exclusion and hierarchy. It appears that discrimination against women in modernized societies is not a residue of backward traditions, soon to fade away. Rather, typical modern ideals of social progress seem to require the structural oppression of women.

To be sure, some women have obtained some benefits, though often only much later, during each of these historical transformations. To question the standards for declaring an era or set of events progressive is not to deny such facts. Rather, the point is to note that in the context of existing hierarchies, it is unlikely that social transformations that already-advantaged groups regard as progressive will equally benefit less-advantaged groups. Even when the disadvantaged struggle to obtain a fair share of the benefits of such changes, it is extremely rare that they have or can obtain access to the resources necessary to succeed. (At least it is rare as a consequence of such struggles. Sometimes unforeseen benefits arrive serendipitously.) One must examine empirically just what the benefits and costs of social change have been for each worst-off group. And, equally important, one needs to investigate why men can't seem to tolerate equality with women.

Categories of social analysis

Kelly-Gadol's second point was that feminist historians were revealing how one must take sex (as she then named it) to be a significant category of social analysis in order to understand and explain social relations past or present. Implicit is the claim "that women do form a distinctive social group and, second, that the invisibility of this group in traditional history is not to be ascribed to female nature. These notions, which clearly arise out of feminist consciousness, effect another, related change in the conceptual foundations of history by introducing sex as a category of social thought" (SRS 813).[3]

What could that mean? Early feminist work tried to conceptualize the relevant phenomena and meanings of sex/gender by analogy with kinds of categories which had been designed to think about other disadvantaged groups. Though thinking of women as an oppressed class, a minority group, or a caste can reveal important aspects of gender relations, none of these ways of categorizing women proved satisfactory. Rather, it became clear that "all analogies—class, minority group, caste—approximate the position of women, but fail to define it adequately. Women are a category unto themselves" (SRS 814, quoting Gerda Lerner). They are "the social opposite, not of a class, a caste, or of a majority, since we are a majority, but of a sex: men. We are a sex, and categorization by gender no longer implies a mothering role and sub-ordination to men, except as a social role and relation recognized as such, as socially constructed and socially imposed" (SRS 814). In making "sex a category as fundamental to our analysis of the social order as other classifications, such as class and race," Kelly-Gadol argued, women's history has made another major contribution to "the theory and practice of history in general" (SRS 816). In chapter 4 we noted how gender (Kelly-Gadol's "sex") functions as a property not only of individuals, but also of social structures and symbolic systems. In all three senses, feminist analyses have expanded the domain of "the social" and "the political" to draw attention to gender differences, a point to which we return.

The sex/gender difference categories have shaped modernization policies, practices, and theories through a series of gendered binaries. Masculine vs. feminine associations enliven and give meaning to virtually all of the central modernity categories. Most conspicuously, as we will see shortly, they give distinctive meanings to future vs. past and public vs. private.

Theories of social change

Kelly-Gadol pointed out that if women are indeed a distinctive social group, namely, a "sex," and if their invisibility in conventional histories is not a consequence of female nature, then "our conception of historical change itself, as change in the social order, is broadened to include changes in the relation of the sexes" (SRS 816). Relations between women and men have effects on history no less than economic, political, or other social relations, she is arguing. Indeed, they *are* economic, political, and social relations (SRS 817). Subsequently, attention to this "intersectionality" of gender with such other economic, social, and cultural features as race and colonialism has led to directives always to examine not only how femininity or masculinity is instituted differently in different social contexts but also how race, class, and other significant social differences are instituted differently for men and women in any and every particular context (cf. Collins, *Black Feminist Thought*; Crenshaw et al., *Critical Race Theory Reader*).

Kelly-Gadol went on to point to the power women lose when public and domestic spheres are separated from each other. Since women are central to the creation and maintenance of kinship relations, they tend to have more economic and thus also political power in social relations structured by kinship, as in aristocracies, tribal social forms, or agrarian societies of family farms. "When familial activities coincide with public or social ones, the status of women is comparable or even superior to that of men" (SRS 818). It is only when public and private spheres are separated that "patriarchy, in short, is at home at home. The private family is its proper domain" (SRS 821). Of course this functional separation of public and private spheres is one of the significant marks of modernity's social progress "for humanity," according to both the modernization theorists and their critics, the dependency theorists, as we will see shortly.

Influential studies by Angela Davis (*Black Women's Role*) and Mina Davis Caulfield ("Imperialism, the Family, and Cultures of Resistance") pointed to the power of the family labor of oppressed peoples in any context of resistance. This labor remains to at least some extent out of reach of the conquering powers—it is the last kind of labor the occupiers or imperialists can directly control. Thus Davis showed the power of women in the family in the U.S. slavery system, where only African American women were available to care for African American children, the sick, and the elderly. Moreover, slave rebellions were planned "around the kitchen table," thus locating political strategizing and organizing precisely in the supposedly private sphere of African American women's worlds. Davis pointed out that the family, which middle-class white women rightly were finding oppressive, was for African American slaves a source of power, status, and creative energy. She implied that in the still racist society in which we live today, African American family labor remains an important site of resistance to white supremacist policies and practices. Caulfield documented this phenomenon in contexts of imperialism and colonialism more generally in her influential study of the importance of women's work in resistance to imperialism and colonialism. Yet national liberation struggles, whether in the West or in the formerly colonized societies of the Third World, frequently result in the formal liberation of men's democratic rights only. Women's destiny after such struggles is all too often to return to the household and kitchen, which, after the revolution, cannot confer the status and power to women that they earlier did. Economic and political issues are inextricably intertwined in these accounts. We return to these issues in the next chapter.

Theories of social change which do not account for how gender relations interact with other kinds of social relations fail to explain what happens in the historical eras on which they focus. Moreover, there are gendered psychosocial dimensions to social change which the modernity theorists rarely address. Kelly-Gadol already saw the importance of the psychosocial when she concluded that "if the historical conception of civilization can be shown to include the psychosocial functions of the family, then with that understanding we can insist that any reconstruction of society along just lines incorporate reconstruction of the family—all kinds of collective and private families, and all of them functioning, not as property relations, but as personal relations among freely associating people" (SRS 823). This influential essay was written three decades ago—a generation. Its lessons have yet to be learned by the mainstream of social theory and, in particular, by both the defenders and critics of modernization theory.[4]

A caution: One might be tempted to interpret Kelly-Gadol as saying that women have been absent from projects claimed to be progressive, such as modernization, in the case here. However, it is worth emphasizing that her argument is a different one: that women have lost status and prestige precisely *because* of the achievements claimed to make the event or era progressive, and that historians have failed to recognize this phenomenon. Women appear as missing in action in projects aimed to bring social progress, although careful observation reveals that the action would be impossible without their loss of status and prestige. This is an important distinction. Shortly we will ask how the apparent absence of women from the public spheres of modernity and modernization projects has in fact masked the preoccupation of these projects with the purported progressiveness of men's escape from, and subsequent control of, women and whatever is associated with the feminine. How has fear of women, femininity, and women's worlds in fact haunted modernity in its political, economic, and social goals? Here we have to turn to the issue of gender coding—of the hidden narratives which structure mainstream thinking about modernity and modernization. This gender coding also structures differences between the West and its Others in modern thinking and practice.

But first let us look briefly at the strengths and limitations of two important feminist philosophic criticisms of the ways tradition and modernity have been conceptualized and at the usefulness of a third approach. Then we will return to the faulty narratives of periodization, the too-limited categories of the social, and the false theories of social change which structure the mainstream philosophy of science thinking about modernity and tradition no less than they do that of the social sciences.

2. THREE FEMINIST CRITICAL APPROACHES TO TRADITIONAL AND MODERN SCIENTIFIC KNOWLEDGE

On the one hand, Northern feminists have pointed to women's active development and preservation of "folk" health, medical, and environmental knowledge in households—that is, outside scientific institutions within our own societies. Women seem to play a more important role in creating and disseminating this kind of "indigenous knowledge" than they do in producing modern scientific and technological knowledge.[5] Around the globe, it is often, or even usually, women who develop and preserve local pharmacologically useful plants and the knowledge of how to apply them, as well as more general health practices and medical treatments. Women have developed agricultural and cooking practices, as well as weaving, pottery making, and all of the associated technologies for such manufacturing. We have argued that these kinds of traditional knowledge, as well as traditional environmental knowledge, are also important forms of empirical knowledge. One early development of standpoint epistemology specifically grounded its arguments in the reliability of women's (culturally mediated) experience of our own bodies—our traditional or "indigenous" knowledge of our bodily processes, one could say. This reliability (but not, of course, infallibility) contrasted with that of the dominant medical and health sciences which had been created by male medical and health professionals who lacked such experience and the knowledge it could generate. Traces of this insight can be found in Hartsock's and Smith's accounts also (Hartsock, "Feminist Standpoint"; Smith, *Everyday World*; Rose, "Hand, Brain, and Heart"; cf. also Martin, *Woman in the Body*). So, one could say, this strategy has sought to justify women's traditional knowledge as a kind of "real science." In doing so, it contributes to the ongoing process of expanding what gets to count as the production of scientific knowledge (Beck, *Reinvention of Politics*; Nowotny et al., *Re-Thinking Science*).

Yet this strategy has not been convincing to those who consider modern sciences such as physics, chemistry, engineering, and biology as different in kind from traditional knowledge, and who think that very difference is what certifies the promise of modern sciences to deliver social progress. Is it different in kind?

A second strategy points to the historic fingerprints on the purportedly transcultural standards of objectivity and rationality for which modern science was to provide the most perfect exemplar. Rationality and objectivity have persistently been associated with masculinity, even as standards for these concepts have themselves shifted over time (Bordo, *Flight to Objectivity*; Jaggar, "Love and Knowledge"; Keller, *Reflections on Gender*; Lloyd, *Man of Reason*; MacKinnon, "Feminism, Marxism, Method"). This feminist strategy can also draw on the valuable critical resources of poststructuralism, which delineates additional features of modernity's delusions of historical and cultural transcendence (Flax, *Thinking Fragments*; Haraway "Situated Knowledges"). So one can conclude that modern scientific work is not culturally neutral and could not in principle attain such neutrality as long as it retains distinctively androcentric and Eurocentric conceptions of rationality and objectivity. Yet that neutrality and the methodological practices designed to attain it were supposed to be the significant differences between modern and traditional knowledge projects.

This second strategy leaves many philosophers and most scientists cold, since they cannot see how such associations, interpretations, or meanings (for example, of the masculinity of modern Western rationality), regrettable as they might be, have effects on contemporary research methods or results. They cannot—or refuse to—recognize that such meanings shape scientific practices, and the content of the claims so produced. Yet such associations in practice affect how scientists conceptualize and interact with nature's order. However, to the philosophers and scientists, scientific work can still retain its transcultural value-neutrality regardless of these cultural meanings or interpretations of it. What is scientific about modern sciences does not include whatever meanings or interpretations people in a particular historical or cultural context assign to scientific methods or the facts it produces. Culture is conventionally conceptualized as an obstacle over which scientific method and its production of facts must triumph. It is precisely these methods, facts, and the prediction and control of nature's regularities they enable that certify scientific claims, according to the conventional view. Science shows that nature's regularities will have their effects on us regardless if one is a woman or man, Hindu, Muslim, or Christian, they argue.

Of course, some four decades of sociology, history, and ethnography of modern sciences have again and again identified precisely how cultural meanings of nature and scientific practice have shaped what and how Western sciences come to know (Biagioli, *Science Studies Reader*).

Yet most of these science studies have tended to a limited conception of the social, avoiding the most controversial aspects of Western social relations in which modern sciences have been deeply implicated, such as empire, colonialism, male supremacy, and the systematic abandonment of concern for the flourishing of whatever modernity defines as traditional worlds. Such issues have been left for the disvalued fringes of science studies, represented by feminist and postcolonial accounts and by critical accounts of the alarmingly tight fit between the projects of modern sciences, on the one hand, and of national security and capitalist expansion, on the other.[6]

Yet there is a third strategy which feminist science studies and its philosophic projects have left undone. This is the critical exposure of the gendered meanings of the modernity vs. tradition binary itself. It is not just that modernity ignores how women make important contributions to traditional knowledge and that modern rationality and objectivity are associated with models of masculinity. In this third case the focus is on how the nature and desirability of modernity and its knowledge systems are, in modernity's narratives, conceptualized from the start in terms of their distance from women and the feminine (as well as their distance from the primitive), signified by what modernity refers to as tradition.

Modernity is not a thing which can be understood or explained by examining it alone. Rather it is always half of a relationship. Like masculinity's relation to femininity (Flax, *Thinking Fragments*), modernity always defines itself as not its Other—not tradition. Tradition is always represented as feminine, primitive, and in modernity's past. Modernity is obsessively preoccupied with contrasting itself and its distinctive features with these Others; the feminine and the primitive always appear in modernity's narratives as the negatives to modernity's positives. According to this account, any traces of tradition which remain in modern societies, such as the oppression of women, for example, are unfortunate residues which will wither away as modernity more thoroughly disseminates throughout social relations.

Note that this kind of interrogation of tradition and its relation to modernity could not occur until postcolonial scholars had demonstrated that the West did not and will not ever have a corner on modernity; modernization is not the same as Westernization. That is, an interrogation of tradition to match the feminist and poststructuralist interrogation of modernity becomes possible only with the recognition that many other societies around the world have developed their own forms of modernity.[7] They have done so self-consciously *within* cultural beliefs and practices that the modern West regards as traditional. Here, postcolonial science and technology scholars have led the way in their discussions of multiple modern science traditions. Thus, for example, overtly Hindu and Islamic *modern* sciences and their philosophies have been developed (Prakash, *Another Reason*; Sardar, "Islamic Science"). Moreover, such studies also reveal the many ways in which modernity reproduces tradition and depends upon it for its own successes, contrary to modernity's claims that tradition is always only an obstacle to its successes and that modernity is completely incompatible with tradition—a point to which we will return.

3. OBSCURED PATERNAL NARRATIVES OF MODERNIZATION

The modernization theory originating in the nineteenth century was reinvigorated by post–World War II sociologists, such as Alex Inkeles and David H. Smith (*Becoming Modern*) and W. W. Rostow (*Stages of Economic Growth*), who were concerned with understanding and justifying the economic, political, and social changes occurring as the West's economic resources were being redirected from war efforts to social transformations around the globe and to Cold War dynamics. Its most powerful critics in the West have been the Marxian-inspired world systems theorists and dependency theorists, such as Wallerstein (*Modern World-System*) and Frank (*Capitalism and Underdevelopment*). In effect they propose an alternative theory of modernization.

In this section I will identify four themes in modernization narratives which reveal their foundations in gender stereotypes, and how the benefits of Western sciences and technologies consequently are delivered primarily within male-supremacist conceptual frameworks.[8] The following section turns to identify some silences and gaps in logic exhibited in these representations of scientific rationality and technical expertise.

Men must separate from their pasts

One theme is that if men would become active agents of their own histories and of social progress in general, they must escape the pull of women, households, and everything associated with them. Modernity's normative power is always carried by the struggle to separate from the past—from childhood; family; the emotions which attach one to childhood, kin, and the past; nature; the pre-modern; tradition; and whatever else is metaphorically conceptualized as women's worlds. That is, it is precisely this struggle which marks modernity as progressive.

Economics and politics must move to the public sphere

A second theme is that political and economic activities must be removed from the private realm of households and exercised in the public sphere, which is defined as men's space. Indeed, this disarticulation of social institutions is always cited as a significant mark of modernity and its social process (Eisenstadt, *Multiple Modernities*). Yet we saw that for women it is precisely their exclusion from these now-public activities which insures their lesser social status and power (Kelly-Gadol, SRS). Hence the continuing importance of women's efforts to gain access to public institutions and, especially, their governing positions. And one can understand, too, why families then provide the sites of greatest resistance to external economic, political, and social intrusions and invasions.

The time machine: Scientific rationality and technical expertise

A third theme in the modernization accounts is that it is scientific rationality and technical expertise that can and must direct economic and political activities in the public sphere. It is their scientific rationality and technical expertise that make economics and politics modern, that transport these institutions into modernity like a one-way time machine of science fiction. Thus it is boys and men who receive scientific and technological training since they are the legitimate residents of the public sphere. Moreover, science and technology are overtly "mission directed" in modernization contexts. However devotedly scientists and engineers conceive the purity of their search for knowledge, it has become clear that the only rational justification for providing the huge public material and social resources required for modern scientific and technological research is to advance economic and political projects, including military ones, in the name of supposedly universal social progress (Forman, "Behind Quantum Electronics").

To be sure, scientific rationality and technical expertise have been deployed in the private sector. Male experts appropriated control of obstetrics and gynecology from midwives. Such rationality and expertise have even been deployed by the great male chefs and the male authors of books on child rearing and by "how-to" authors on the topic of sex. Moreover, in the early part of the twentieth century, some socialist feminists developed "technocratic feminism," which tried to rationalize the design of homes and the nature of domestic activities.[9] Yet this infusion of public-sector rationality and expertise into the private sector neither eliminated the gender stereotypes which structured the modern division of social life into public vs. private spheres, nor shortened the hours women were expected to spend on traditional domestic labor.

Social progress through evolution

Finally, it is an evolutionary model of social and political change which governs these transformations. Development was conceptualized by the modernization theorists as the struggle for maturity and thus the achievement of dominance of men and their modernity over nature and women. "In using an evolutionary model, they portray development as the ever-widening ability of men to create and transform their environment" (C. V. Scott, G D 24). This was to be accomplished through political and economic policies shaped by scientific and technological practices. These would "leave women behind" in the private and natural world of the household. "Women's continued subordination in fact defines male citizenship" (24).

Moreover, throughout this work, "the comparison of the liberated and independent woman of the West with the tradition-bound woman of the Third World also informs many accounts of the psychosocial requisites of modernity" (G D 25). Third World women are presented as "uniformly oppressed by men and family structure" (25). "Such contrasts not only serve to establish a Western sense of difference and superiority (and complacency about women's rights in the West); they also mark women, in Mohanty's terms, as 'third world (read: ignorant, poor, uneducated, tradition-bound, domestic, family-oriented, victimized, etc.).' As the most 'backward' group in society, women serve as an implicit contrast between Western modernity and non-Western tradition" (26).

Scott goes on to show how such gender stereotypes direct both the modernization policies and practices of the World Bank and also the Marxian theory which set agendas for nation-building policies and practices in the revolutions of Mozambique and Angola. How rational are these assumptions?

4. SILENCES AND GAPS IN LOGIC:
THE QUESTIONS OF OTHERS

Feminist and postcolonial readings identify revealing silences and gaps in the logic of these modernization narratives. While explicit references to women or femininity only rarely appear in these texts, they always specify primitive or traditional features which modernity finds unintelligible, illogical, irrational, or ridiculous. It is women and the feminine which are the primary exemplars of such properties in everyday normal misogyny, as Freud put it. Women and femininity are invoked to emphasize the borders or horizons of the modern—whatever is inconceivable for modernity to think. Thus, when notions of modernity are built upon stereotypes of the masculine, modernity's "Others," regardless of who they are and of the content of that contrast, are always feminized. Let us look more closely at what is wrong with some of the assumptions and claims of modernization theory that lead to such conclusions.

Is modernity incompatible with tradition?
Not at all. To the contrary, modernity requires the continuation of tradition in its very self-definition. Modernity narratives are always about time and historical change. So the past is represented in ways that contrast with the modern; modernity defines itself against whatever it defines as the past. Consequently, modernity narratives obsessively recuperate feminized tradition in order to define their own different, manly, and Western progressive features. In this way, tradition, exemplified within modern societies by women and women's worlds, becomes conceptually internal to modernity.

Moreover, these accounts selectively appropriate traditional features, especially social hierarchies that they encounter, reshape them to suit modernity's goals, and then obsessively reproduce them. Postcolonial critics point out how modernity narratives radically reshape so-called traditions and reposition them in ways that make modernity look desirable. For example, suttee, the Indian widow-burning practice, was restricted to only a small section of India prior to the arrival of the British. It spread in resistance to the British attempts to eliminate Indian traditions as backward and savage (Mohanty, *Feminism Without Borders*; Narayan, *Dislocating Cultures*). Thus those maligned by modernity's agendas can critically reevaluate just what are the desirable traditional features of their own cultures.

Modernity depends on the survival and flourishing of traditional activities in a number of other ways. The new must always be "sutured" into prevailing cultural practices and made acceptable and desirable to groups potentially hostile to it. Murata ("Creativity of Technology") shows the necessity of traditional craft labor to create the infrastructure in Japan for the introduction of such modern Western technologies as railway systems. "Practice theorists" in philosophy, such as Rouse (*Knowledge and Power* and *How Scientific Practices Matter*), have pointed to ways in which the careful labeling and organizing practices of the laboratory have escaped into our kitchens and medicine closets. We can then see that it is mostly women's labor which must create the laboratory conditions in the household in such cases and mediate acceptable social relations between traditional pharmacological or cooking practices, on the one hand, and the new scientific ones, on the other hand. Modern conceptions and beliefs would never come into practical existence without the traditional labors of women and other purportedly pre-modern groups. Thus tradition is neither only in the past nor fixed and static. Once modern tendencies appear in a society, tradition becomes internal to modernity and evolves along with other modern institutions and practices.

Are households to be abandoned by modernity?

Both modernization theorists and their Marxian critics focus on the public realm of production and, for the modernization theorists, Liberal (that is, vs. Marxian) democratic politics as the motor of social progress. Both economic production and democratic politics are to be designed and managed by scientific rationality and technical expertise. It is transformations in the public realm which will bring about social progress for humanity. But who is to be responsible for the flourishing of households? Care of children, family, and kin relations; cooking, shopping, and other household tasks; the "emotional labor" which everyone needs and which keeps us sane in a "crazy world"; and the maintenance of community relations—all these are prerequisites for life itself for those who work as wage laborers and as managers and administrators of social relations of the public sphere. But evidently the flourishing of this "private sphere" is not the responsibility of modernity. Dominant conceptions of modernization preclude social progress for households and the worlds for which women are assigned responsibility. The private sphere is in effect enslaved and exploited by modern public sphere institutions—economic, political, educational, and governmental institutions, as well as their policies and practices.

In both modernization theory and the Marxian alternatives, the continued domination and exploitation of women in modern societies are persistently claimed to be an unfortunate residue of pre-modern beliefs and practices which will disappear when women can participate in wage labor and democratic politics in the public realm. An influential early study of the successes and failures of socialism in Czechoslovakia pointed out that while socialism delivered many benefits to women, it ultimately failed to liberate them and thereby weakened its other projects because it could not—would not—deal with social relations between the genders in households (H. Scott, *Does Socialism Liberate Women?*).

Even apart from issues of child care and maintenance of kin networks, one can spot in the Marxian accounts a mysterious silence about the realm of "consumption." The success of production depends upon the successful organization of consumption of the goods and services which are produced. Without appropriately managed consumption, production fails as gluts of products pile up unbought and unused or shortages of goods and services increase. Consumption includes not only shopping but also such tasks as turning the raw fish, yams, and artichokes into edible and culturally desirable foods. What would our understanding of economic relations look like if one started to think about them from the standpoint of the lives of those who do the labor of organizing and processing already produced goods and services for consumption instead of from the standpoint of the lives of those who just produce?[10]

Today in the West, armies of largely immigrant women and men from the Third World take care of the children, households, and gardens of professional and other working women and men and provide services in restaurants, hospitals, schools, and offices; in pre-modern cultures they were part of traditional labor (cf. Sassen, *Globalization and Its Discontents*). These workers are often also maintaining their own families, households, kin networks, and communities in the Philippines, Latin America, and elsewhere as they work for the technical elite classes in the North (and South). I return below to reflect further on the emergence in globalization of these kinds of globally distributed households.

Are modern men autonomous and themselves responsible for their own achievements?

Of course not, as a generation of feminist accounts have detailed. Their ability to engage in managerial, administrative, and professional labor requires the armies of workers tending their bodies, their families, and the local places where they work and live (see the section "Is modernity incompatible with tradition?"). Modern men are freed from having to tend their bodies by the body-work everyone else must do. These others are mostly but not entirely women (cf. Smith, *Everyday World*).

Is the modern model of the ideal relation between speech and authority (the narrative of rationality and expertise) suitable for human social progress?
No. This narrative of rationality and expertise recommended by and characteristic of modernity is a parochial monologue serving only the interests of the powerful (Harding, *Science and Social Inequality*, chap. 7). As legions of feminist, postcolonial, and other critics have argued, it lacks the critical resources to grasp its own irrationality, the limits of its expertise, and thus its own parochiality.[11] Whatever may have been the case in the past, today a society cannot reasonably be regarded as flourishing when the vast majority of its citizens and millions of citizens of other societies must lose social status and power over their own lives in order that a small minority gain even greater status and power. Prevailing philosophies of science advance standards of authority grounded in conceptions of their objectivity and rationality; yet these conceptions lack empirical support, as noted in earlier chapters. Their continued circulation dupes their believers into accepting the dominant false and self-serving accounts.

Why do gender and race/imperial discourses mutually circulate in modernization theories?
In short, each needs the other for its own successes. As the critics point out, those horizons that modernity creates to cordon itself off from its now unintelligible past are always both patriarchal and racial/colonial/imperial. Women are consistently represented as outside history and society—outside "civilization" (cf. Wolf, *Europe and the People Without a History*). They are primitive, incomplete, immature forms of the human. They appear as an ahistorical Other, like the noble (or sometimes ignoble) savage. Women represent nature, tradition, the emotions, as well as the mother's body. "The redemptive maternal body constitutes the ahistorical other and the other of history against which modern identity is defined" (Felski, *Gender of Modernity* 38). Consequently it should come as no surprise to discover that actually and supposedly pre-modern peoples and practices are themselves coded feminine in

modernity narratives (Stepan, "Race and Gender"). Women are like "savages" and "savages" are like women in such accounts. Hence attempts to transform only one of these two kinds of hierarchies are doomed to fail, since each can continue to exercise its powers "inside" the Other. Thus gender and science projects cannot succeed in eliminating gender hierarchy in the sciences as long as the West vs. Rest hierarchy thrives, and vice versa.

On the one hand, this recognition can be discouraging. On the other hand, it reveals the excellent positions that feminist and postcolonialism/anti-racism projects already occupy when they conjoin in coalitions to strategize how to achieve kinds of social transformations for which both yearn. Only such political movements that take on multiple oppressions will in fact be successful. This historical moment of the "crisis of the West" and its simultaneous "crisis of masculinity" provides incomparable opportunities for social transformation that other falsely regarded progressive moments lacked when they were focused on only one kind of pattern of social discrimination.

Why does modernity's narrative replicate the Freudian narrative?
Finally there is one small point—to be sure, controversial—which must at least be mentioned. There is an eerie echo in both the modernization theories and their Marxian alternatives of Freud's narrative of how a boy becomes a man. Numerous commentators point to the startling fit between these narratives of modernization and the standard psychoanalytic-influenced narratives of child development. The three main resources for the arguments of this section all point out this phenomenon (Felski, *Gender of Modernity*; Jardine, *Gynesis*; C. V. Scott, G D). It is the "mother world" against which children must struggle in order to separate and individuate into mature children of one culture or another. They must learn to control their natural bodily processes (that is, nature). In the Freudian and the modernization narratives, maturity requires a painful struggle against the women-organized world of kin and household. Achievement of rational adulthood separated from women's worlds, and linked with the control of the natural world, is the goal of both.[12]

To be sure, science studies and philosophy of science are probably just about as hostile environments as one could find in which to raise such an issue. Of course, there are plenty of limitations and problems with such psychoanalytic accounts. Yet I suggest that there are also good reasons to pay attention to them. What should we make of the fact that the narratives of modernization seem to replay these narratives of child development?

One reason to attend to them is that the psychoanalytic accounts direct attention to anxieties and fears that the modernity and Marxian theorists themselves find reasonable and, moreover, expect to be compelling to their audiences. The theorists do not much censor the gender-coding themes in their narratives. Such associations of modernity with manliness and tradition with femininity do not seem odd or inappropriate to the modernization theorists or their Marxian critics.[13] At one level they probably are not even aware of them since such gendered stereotypes are not invented by them but rather prevalent in their culture, as well as presumably part of their collective and individual psychic structures. Feminist criticisms of gender stereotypes, though these already appeared during the nineteenth-century women's movements (think of Sojourner Truth's challenges to bourgeois norms of femininity in the mid-nineteenth century), were not part of these theorists' cultures. So even if they were aware of such associations, they would probably not have seen anything wrong with them.

A second point is the suggestion of feminist psychoanalytic accounts that equitable social relations between the genders will remain an unsuccessful struggle for both women and men as long as the primary responsibility for child care is assigned to women and that the direction of public life is assigned to men. The constitution of gendered public and private spheres, whatever its psychic origins, can support only gender inequities, as the feminist critics of modernization theory consistently argue. So if adult gender relations through which children's gender identities are formed do structure those children's adult fears and desires in ways which support social inequalities, universal social progress requires that the world of social relations within which children's gender formation occurs be addressed.

Finally, a third point is the possible light thrown on Freudian theory by its apparent close relation to Western modernization theories and practices. Critics have always complained that Freudian child development theory does not hold for non-Western societies. Perhaps its relation to modernization in part explains such a limitation. Perhaps Freud, too, should be included in the pantheon of great, flawed nineteenth-century modernization theorists.

5. CONCLUSION

My major point in this chapter has been that the widespread prevalence of gender stereotypes within modernized societies is not a mere residue of traditional social relations, as modernization theorists and their Marxian critics have consistently proclaimed. Rather, these stereotypes are built into the founding conceptual framework of modernization thinking; it is stereotypes of the masculine, defined in terms of their distance from "the feminine," that indicate what counts as modernity and what counts as social progress. Moreover, scientific rationality and technical expertise are conceptualized as the motor of modernization as well as of masculine achievement; like science fiction time machines, they transport men away from loyalties to women's worlds into modernity's socially progressive worlds of economic production and democratic government, both of which have been removed from the polluting direction and management of women.

This is not to say that individual "modern men" make no effort to enable their wives, parents, children, and local communities to flourish. They often do, and at great personal cost.[14] Rather, the point here is that the institutions and practices in their worlds defined as "modern" treat women, households, children, kin, and local communities as, at best, merely exploitable resources to advance public-sphere economic, political, and social goals. The connections men feel with these traditional worlds are regarded as obstacles to their personal achievement of autonomy and rationality and to their society's attainment of modernity. What modern sciences and technical expertise do is "externalize" men from women's lives and households. At the same time, these narratives obscure how the achievements of "individual" modern men and those of their modern institutions more generally always remain dependent upon women's so-called traditional activities of maintaining responsibility for children, households, kin relations, and the flourishing of the communities and the natural environments upon which such activities depend.

Are men merely optional to women's social progress? Insofar as modernity's ideals unrelentingly define social progress in terms of escaping and abandoning women, women's worlds, and anything associated with them, such ideals would seem to leave men in their official, public capacities as irrelevant to women's social progress. The feminist analyses here make it clear that it must be women who define what should count as social progress for themselves and for their children, kin, households, and communities. Men can join such projects or not, but it is clear that it would be unwise to expect the public-sphere institutions which demand men's escape from household responsibilities, or the men who think such demands reasonable, to be able to design and manage truly progressive social transformations which could deliver the social benefits women need and desire. Clearly, progressive social movements—no less than the mainstream institutions which have been the main targets of these criticisms—need to rethink this issue.

Of course much of this argument has long been made by feminist political philosophers and other critics of the public vs. private spheres of Liberal democratic theory. What is different here is the exploration of how this kind of argument can and must be directed to the modernity vs. tradition contrast and to the roles philosophies of modern sciences and technologies play in maintaining this contrast. Philosophies of modern sciences and technologies are much more deeply implicated in male-supremacist theories and practices of modern social life in these arguments.

What is to be done about the way gender stereotypes—past, present, and possibly future—give content, meaning, and moral energy to prevailing conceptions of modernity, tradition, and social progress?

9.

MOVING ON

A Methodological Provocation

— — — —

THIS BOOK BEGAN WITH THE SUSPICION that the widespread use of the contrast between modernity and tradition has functioned to hide a number of problems with standard Western ways of thinking about social progress, and especially about possible scientific and technological contributions to such ways of thinking. This binary has long structured the conceptual frameworks of the natural and social sciences in the West, and often also of the work of non-Western thinkers. One can now see that exceptionalist and triumphalist evaluations of Western modernity and its modernization projects in the Third World have obscured the persistence of legacies of male supremacy and Eurocentrism in the work of even progressive science and technology scholars and activists.[1] The persistence of such assumptions in this work is discouraging since so much of their effort has seemed to promise otherwise hopeful transformations of both the sciences and the domain of the political. Yet now we can wonder if any such transformations can have positive effects on the lives of the vast majority of the world's citizens as long as the latter remain represented, explicitly or implicitly, as the disvalued Others against which the success of progressive social transformations are measured. Of course this is not to say that such scholars have intended to promote male supremacy and Eurocentric imperialism when strategizing about how to advance social progress, but only that their work has so functioned because of the insidious male-supremacist, racist, and imperial effects of the contrast between modernity and tradition.

Thus the earlier chapters have been calibrating to each other the potentially politically progressive aspects of scholarly research projects and activist concerns which are not usually visible at the same time. A familiar trope recommends keeping "both eyes open" to achieve the valuable "double vision" that comes from calibrating a dominant group's favored historical account to the realities experienced by those they have dominated.[2] However, we have been building up, chapter by chapter, the much more difficult task of trying to keep simultaneously in view some five different kinds of research agendas which do not much include each other's concerns. To use another metaphor (and to avoid having to think about keeping five "eyes open"), some five or more intellectual and political "plots" must be kept "on the stage" of our thinking at once. These include mainstream assumptions about modernity and its scientific and technological projects (including mainstream progressive post-Kuhnian science studies); postcolonial science studies; feminist science studies, both North and South; historians' research on multiple modernities; and feminist critiques of modernity.[3] It turns out that the same kinds of historical agents or "actors" appear in each drama, some strutting at center stage and others lurking or cowering in the background. Looking at each plot from the perspective of the others has enabled us to identify how important obstacles to progressive transformations of science and politics are to be found in the overlapping roles that occur among these dramas. Western science's "modest witness," already identified in science studies as the emerging democratic citizen (Shapin and Schaffer, *Leviathan*), turns out also to be the young hero fleeing the ties of his childhood in the world of mother, kin, and village for his autonomous, rational adulthood in the modern city. The "tradition" he sought to leave behind is exemplified by his mother, his fiancee, and his servants or rural peasants—or, in urban centers today, his wife's world of women, children, and household management and the worlds of the migrant labor producing his food, manufactured goods, and daily services.

So what is to be done? What should those with more comprehensively progressive intentions do—that is, those who refuse to join the project of leaving behind women and non-Western men as reformers seek to transform sciences and the domain of the political in directions which it seems will primarily benefit persons like themselves? Of course we can't go back to live in the pre-modern, a point to which I will return in section 3. Yet if we could, those worlds were not necessarily completely wonderful even for dominant groups of men, let alone for women or devalued groups of men; men as well as women and other devalued groups have good reasons to want many of the benefits that modernity's social progress clearly can deliver. However, it will not be sufficient simply to include women and non-Western men in existing progressive science and technology projects, important as such projects are, since for dominant groups the very definition of progressive leaves them conceptually dedicated to leaving behind women, the poor, and their dependents—that is, the vast majority of the world's least-advantaged citizens. Furthermore, while direct confrontation of dominant institutions by those they wrong does indeed have its benefits, it also reproduces the very binary categories which keep such social hierarchies relevant and functioning. It is hard to see a way forward when the very notion of social progress itself turns out to be discriminatory against the majority of the world's citizens, often including precisely the groups who are supposed to benefit by such progress. This is the problem which this chapter begins to address.[4]

First, a brief summary of the resources Northern science and technology studies and the studies from modernity's peripheries can contribute to each others' projects. Then, a provocation: a proposal for a possible methodological response to the problem of how to move forward without further advancing the regressive male-supremacist and imperial projects which lie deep in modernity's conception of progress. This proposal will not resolve all the problems with the modernity vs. tradition contrast, but it can provide a helpful shift of direction in strategies for addressing them.

1. RESOURCES FROM THE NORTH

From representatives of three different foci in the field of Northern science studies, we began to see that modernity's periodization scheme, its model agents of knowledge and history, and its theories of social change lack empirical support.[5] Bruno Latour pointed out that "we have never been modern," since in everyday and public life no one finds it useful to separate facts from values in the kinds of ways demanded by the modernity ideal. It is scientists as a group (and their philosophers) and the politicians who rely on them who have gained an illegitimate authority about both the nature of the world and desirable social politics, through their insistence on a fact/value distinction which they can't even achieve in their own best work. Latour traces this problem back to Plato's Myth of the Cave and tries to envision a different kind of conceptual framework for sciences and political relations in an "amodern" society, one which retains valuable functions of the fact/value distinction but jettisons its problematic features.

Latour's arguments represent a powerful, though apparently largely inadvertent, leveling of the playing field between representations of nature characteristic of the West and representations characteristic of its Others. The power of the sciences of the West has depended in part on contrasting the purported progressiveness and cultural neutrality of Western sciences with the anthropomorphizing of nature, which is supposedly characteristic only of non-Western cultures, and with women's supposedly only subjective perceptions and evaluations. During the last three decades, Latour has again and again made important contributions to Northern science studies' arguments that sciences and their societies in the West have consistently co-constituted each other. Western sciences are always situated and local, exhibiting historical and cultural "integrity" with their particular eras, which they in turn help to constitute and direct. We, too, anthropomorphize the natural world just as pre-modern societies do, and we do so in our very best scientific and technological achievements, not just in our worst. Thus the conventional fact vs. value foundation of the modern vs. traditional contrast cannot find empirical support in the history of Western sciences, contrary to the assumptions of conventional philosophies of science and of modernization theories. And modernity's periodization scheme is erroneous, since central features of modernity seem never to have actually arrived in societies regarded as modern. For example, as we discover elsewhere, gender hierarchies do not disappear, as they should if the modernization theorists were right. Rather, modern tendencies transform them to serve modern projects, and in different ways in different cultures. Latour's argument has an additional virtue for our project. He refuses to presume that one must adopt a postmodern stance if one criticizes the modernity we have. He opens the door to imagining alternatives which have not been visible in the modernity vs. postmodernity debates.

Ulrich Beck, the "risk society" sociologist, interrogates modernity and its sciences from a different set of concerns and from a rich background in social theory. Like Latour, Beck shows how scientists illegitimately appropriated as merely technical decisions matters which should have been publicly debated as political choices. To the scientists, it is inefficient and inappropriate for such decisions to be decided through public debate; to Beck such intrusion of science into the political domain is unjustifiable on both scientific and political grounds. Beck goes on to identify three ways that scientists' monopoly on the production of scientific knowledge has already been slipping away in the emerging "second modernity" on which his work has focused. The sciences of science, as he refers to the social studies of science, reveal aspects of nature/society's order as they critically examine just how scientific knowledge is produced. It is not just official scientists who can and do provide valuable information about the natural/social world around us. Beck also notes that all of us seem forced to participate in the production of scientific knowledge—to become scientists— when having to make decisions in the face of conflicting scientific expertise. Finally, people's daily experiences are producing demands that new questions be addressed by scientific institutions. A kind of "science of questions," as Beck puts it, has widely emerged from everyday life and especially from new social movements such as the environmental movement in which Beck has worked and from global feminisms.

Beck provides another way of challenging modernity's standard periodization scheme. There are two modernities, he argues. First came industrial modernity. But in the present era "reflexive" modernities emerge alongside continuing forms of industrial modernity, thereby shrinking industrial modernity's effective intellectual and political domains. Industrial modernity seems in some respects to be withdrawing from some of its characteristic sites.

In challenging the idea that it is only traditional scientific experts who can advance the knowledge of nature, and thus social progress, Beck also challenges modernity's conceptions of who are the agents of modernity and its social changes. It is not just official experts about nature and social relations who have powerful, and yet still rational, effects upon modernity and its projects. Moreover, he also identifies some surprising aspects of modernity which rarely appear in other accounts. For example, he points out that both industrial first modernity and its reformers in the reflexive second modernity generate counter-modernities of both the "right" and the "left." These counter-modernities will not disappear through confrontation; confrontation with modernity is precisely how both kinds have formed. So, he argues, a different political challenge emerges: how can these counter-modern tendencies be recruited to join progressive elements of the new reflexive modernity? This is a hugely important question which has received far too limited attention, at least in the progressive social movements of the West. To put the issue another way, how can the modern vs. counter-modern binary itself be deconstructed—historicized, complexified, made to seem less inevitable—in ways that promote a more reliable understanding of social progress?

Beck also stresses the necessity of paying attention to the "reflexive" character of scientific research—namely, the way it changes the world it studies, and does so in necessarily unpredictable ways.[6] This aspect of scientific research has been underappreciated in first modernity's philosophies. It requires vastly increased accountability and responsibility for scientific research on the part of scientists and the institutions sponsoring them. For instance, Sandra Steingraber suggests that we could require the institutionalization of a kind of precautionary principle at the heart of scientific protocols (*Living Downstream*). Further thought is required to spell out the various transformations required to enable and mandate such accountability.

In the books by the team of Gibbons, Nowotny, and Scott (GNS), we saw a third approach to rethinking modernity and its sciences, this time from a base in science policy discussions. This team of social scientists focuses on the new forms of producing scientific and technical knowledge which have flourished since 1989, as university-trained scientists have taken up work in industrial and federal government laboratories and, in turn, as university science departments have received funding from industry and government institutions and have been more directed by their intellectual priorities. Like Beck and in a different way Latour, GNS delineate how scientific and technological research institutions, on the one hand, and industrial and government institutions, on the other hand, have both become increasingly transgressive in each other's work and also more permeable by each other.

They point out that this phenomenon seems to indicate the diminishing of the disaggregation of social institutions, which has always served to mark the distinctiveness of (Western) modernity. Is modernity disappearing in this way too?

Moreover, they point out that these new industry/government/science relations undermine central pillars of Western philosophies of science as they advance the growth of knowledge not in spite of but *through* their mission-directedness. GNS argue that the demands for reliability of the results of research are leading scientists to become more responsive to the social, economic, and political contexts of their work. In such contexts it is no longer meaningful to try to keep apart purportedly pure research and its applications, or the internal logic of science and the social contexts of research. Surprisingly, such contextualization of research *strengthens* the reliability of its results; this social robustness increases rather than decreases the reliability of research results, counter to modernity's required "social segregation" standard. Another phenomenon GNS bring to our attention is that modernity and modernization seem to be parting ways. As the legitimacy of modernity's philosophy of science decreases as a result of the kinds of criticisms GNS and others have made, modern technologies have become more powerful. How should we understand this surprising phenomenon?

These three accounts provide useful resources for rethinking the modernity vs. tradition contrast, its effects upon the production of scientific knowledge, and possible strategies for transforming both (Western) sciences and the domain of the political. Precisely because these accounts are, and are perceived to be, closer to mainstream scientific practice and philosophies of science, they offer valuable resources for the social justice movements projects of Part II. They identify contradictory aspects of modernity and its sciences and new directions in the production of scientific knowledge—some promising, some alarming—which can serve as points of entry into mainstream science and technology discourses for the studies from the peripheries of Western modernity. Yet all three, like much of the rest of the field of science studies, lack the kinds of resources which have been emerging at the peripheries of modernity from precisely the interests and desires of those social groups against which modernity has historically defined its projects. All three recognize in different ways the value of feminist work, but do not engage with it. None seem more than marginally familiar with the huge field of postcolonial work, including its science and technology studies. Thus the field of science studies can benefit from resources produced from (Western) modernity's peripheries.

2. RESOURCES FROM (WESTERN) MODERNITY'S PERIPHERIES

Part II brought into focus reasons for these limitations in the Northern accounts. Favored periodization schemes, model agents, and theories of social change have reflected the interests primarily of those who have benefited most from Western conceptions of social progress. In this part of the book we saw women's movements creating new subjects—new speakers—of knowledge and history with distinctive epistemological and methodological tools (to which we return shortly). Postcolonial science studies have also raised new kinds of critical historical and epistemological questions about Northern sciences and about knowledge traditions in the South. Both kinds of questions had been blocked as unintelligible from within the North's modern vs. traditional framework. These studies raised questions about how progressive modernity and its sciences and technologies *ever* could be, insofar as they ignore or trivialize the interests and concerns of the vast majority of the globe's citizens. As long as such interests and desires continue to be represented as beyond the borders of intelligibility for modernity's preferred ways of thinking, damaging exceptionalist and triumphalist tendencies will continue to deteriorate the objectivity, rationality, and social/political accountability of modernity's sciences and their philosophies. Feminist science and technology studies in the South have raised distinctive issues informed by their own positions in postcolonial histories, as well as by recent discussions of these histories and of continuing Western imperial policies and practices. The postcolonial work produced five proposals for future possible relations between the science and technology traditions of non-Western societies and those of the West. Most likely, not one of these will come to dominate global knowledge production; rather, aspects of all five will be evident around the globe in the future.

Part III has directly addressed the modernity vs. tradition contrast from two directions. First, it is now clear that modernity must be plural. Postcolonialism's multiple sciences project cleared the path for this understanding, though there is surprisingly little focus on these science issues in the multiple modernities literature when it is generated from the North.[7] Here we see not just the fact but the necessity of multiple development of modernities and their sciences. The latter must always be "sutured" into local material, social, political, and cultural contexts in order to work at all. Thus a purportedly universal modernity can never completely replace local traditions; it must always (selectively) appropriate and redirect them to serve its own purposes. Moreover, modernity itself must always reproduce something it calls tradition as its Other in order to come into existence in the first place and to remain attractive in the face of deep psychic yearnings for the past.

Chapter 8 began another kind of reframing of the modernity vs. tradition contrast. It showed how this contrast has been haunted by ghosts of the feminine and the primitive in Western, Liberal modernization theory and in the World Bank policies guided by it, as well as in post-Marxian dependency theory and its guidance of nation-building projects in Africa. Here one can observe how modernity appropriates the traditions it finds and generates new "traditions" to serve the needs and desires of the idealized masculine models of modernity which populate these theories. Clearly the nature and purposes of such invocations of "tradition" need a kind of interrogation they have not had and could not get within the conceptual frameworks of modernization theories, Liberal or Marxian, and the institutional policies and practices they guide. It appears that the "crisis of the West" so widely noted in postmodern and postcolonial writings is also a "crisis in masculinity" which is of global proportions.

3. ADRIFT IN A SEA OF UNAPPEALING CHOICES?

The litany of criticisms above and in earlier chapters can be discouraging and overwhelming on first encounter. This is especially so when the conceptual frameworks of even progressive scholars and activists have come to seem less plausible and less desirable than had been imagined. What is to be done?

The first thing to recognize is that a number of apparently obvious choices should be unappealing. For example, we can't, and shouldn't want to, go back to the pre-modern or to "tradition," in spite of Latour's argument that we have never left it. It might seem that returning to the traditional was the point of chapter 8, since from the perspective of the feminist arguments given there it seemed so wrong to base claims for social progress on the abandonment and rejection of women, the feminine, and whatever else is associated with tradition, such as nature, loyalty to kin and community, and respect for the past. Certainly women, who have struggled so vigorously to enter modernity's public spheres as equals, should not be shoved back into households, losing whatever power and status in wage-labor and political life they have gained. But a return to any kind of kin-based society, even if these did give higher status to women, cannot be a progressive desire, even if it were a possible strategy. As already indicated, "traditional" and pre-modern societies have not always been that great for women and other devalued groups who have resided in them, or even for the dominant groups. Non-Westerners and women in the West have very good reasons for wanting a fair share of the many benefits that modern societies and their scientific rationality and technical experts can deliver.

Nor can we "return to the modern"—or, at least, to the first modernity of Beck or to the policy of "segregationist" institutions that Gibbons, Nowotny, and Scott criticize. We have moved past a historical divide between, on the one hand, the world in which those modernities could command rational empirical and theoretical support and, on the other hand, the world we live in now, in which the shortcomings of those modernities are all too evident. Furthermore, we cannot leap into the embrace of any postmodernism which has no recognizably effective strategy for decreasing the immiseration of the vast majority of the world's citizens who struggle under the burden of the costs of the West's "first modernity" projects.

Last but not least, the answer to this question of what to do is not up to "us" alone, that is, to those of us in "the educated classes" in post-industrialized societies who have most benefited from modernity's projects. That arrogant intellectual and political posture is the first feature of the first modernity which we must make sure to leave behind.

At this point we seem adrift in a sea of unappealing choices. However, this very "crisis of the West," which appears also to be a "crisis of masculinity," creates an exciting moment for rethinking available progressive opportunities. Though it is not from the West or from men alone that cultural energies can emerge to generate valuable new directions in social progress, it would be irresponsible for those of us active in such globe-circling movements as feminisms and postcolonialisms, and especially in their conjoined projects, not to try to participate, with modesty and humility, in making contributions to such work. Since male supremacy and Eurocentric imperial tendencies each remain alive and well in the other's purportedly socially progressive projects, it is activity in the conjunction of these two kinds of movements which appears to offer the only hope of escaping the clutches of the two kinds of contemporary crises.

Indeed, clues from the preceding chapters can provide life preservers to keep us afloat as we seek more solid ground. One is that we don't have to set out to invent such projects from scratch or carry them out alone. Desirable futures are already beginning to unroll before us in helpful directions. The feminist participants in the projects of postcolonial movements and the postcolonial participants in the projects of feminist movements have provided perceptions not only of how scientific and technological traditions around the world actually function, but also of how they can guide socially more progressive policies for both research and politics. We can join the local segments of such movements and participate in our own ways in advancing these projects. We can't be sure that such projects will have only progressive effects, but that kind of uncertainty will be our permanent and useful companion!

Another directive comes from standpoint theory/methodology (chapter 4) and its focus on the way in which "doing" both enables and limits "knowing." That is, how we live and interact with the natural and social worlds around us powerfully shapes us so that we will be in a position to understand those worlds. For example, as teachers and scholars we can explore notions of expertise and authority which do not depend upon the old exceptionalist and triumphalist understandings of modernity, its sciences and technologies, and its constraints on social progress. What kinds of expertise and authority should we claim in the classroom and in our writings and want our students to learn, and what kinds should we abandon? A related issue is to ask, What kinds of (feminist) distinctive expertise can men develop from their daily lives? What kinds of anti-imperial, postcolonial expertise can those of us who have benefited from Eurocentric imperialism develop from our daily lives?[8]

There are additional ways in which how we live can provide opportunities for thinking in fresh ways about modernity and tradition.

Here I turn to one small provocation to begin to address directly how to "exorcise" the specters of the feminine and the primitive that are haunting Western modernity and its notions of social progress. For each proposed science and technology research project or policy decision we could ask this question: What would we learn if we started thinking about it and its effects not from the dominant conceptual frameworks, but rather from the daily lives of those groups forced to live in the shadows of such specters—namely, those who have benefited least from the advance of modernity's so-called social progress? How do those daily lives, as they are lived today, provide opportunities for fresh insights about modernity and tradition?

4. START OFF RESEARCH FROM "WOMEN'S LIVES IN HOUSEHOLDS"

Recollect the feminist standpoint mantra to "start off research and politics from women's lives," rather than from the conceptual frameworks of the research disciplines, to create the kinds of knowledge that women need and want to empower themselves and their dependents— children, kin, households, and communities. This was modeled on the Marxian directive to start off from workers' daily lives to understand how the economy works and to provide resources for workers' efforts to better their lives. We saw the postcolonial scholars and activists starting off from the lives of those ruled by Western imperial and colonial projects in order to understand how imperialism, colonialism, the "voyages of discovery," and Third World development policies had largely horrible effects upon the societies Europeans encountered, and especially how scientific rationality and technical expertise functioned within such projects primarily to benefit Westerners. Standpoint projects are designed to identify, explain, and transform the conceptual and material practices of power of the dominant social institutions, including research disciplines, in ways that benefit those who are least advantaged by such institutions. They start off from the lives of the oppressed, but they do not end there, as, for example, do conventional ethnographies. Their main task is to "study up," to identify and explain the material and conceptual practices of power which are often undetectable by those who engage in them (Harding, *Feminist Standpoint Theory Reader*).

The proposal here is to start off research on any attempt to create "social progress" not from the conceptual frameworks of the dominant institutions of modernization, or even from "women's lives" in general, but rather from how women's lives are organized in households. Here is the test which could usefully be applied to many—perhaps most or even all—social and natural science research projects: What can we learn about the research topic or the policies and practices likely to result from it by starting to think about the project from the standpoint of women's lives in households? After all, it is in households that human life is reproduced in significant ways, along with the daily conditions necessary for its continued flourishing throughout human life cycles. (Of course, households are not the only places where human life is reproduced.) I am not suggesting that this is the only useful social location from which to start off progressive research; rather, I am suggesting it is one which any research project seeking to advance the growth of knowledge and social justice should undertake at some point. It is a prerequisite for maximizing validity and objectivity, as well as social justice.

Of course such a methodological directive will be controversial, a point to which I will turn shortly. Yet I suggest that it will be valuable for several related reasons. First, it brings to a focus much of what the modernity and modernization theory has characterized as obstacles to its own progressiveness. It starts off thinking about modernization, tradition, the global political economy, and other topics precisely from the standpoint of those human activities that are the most disvalued in the modernity narratives and models of social progress. Indeed, the household is "where patriarchy is at home," as several feminist critics have pointed out (Jardine, *Gynesis*; Felski, *Gender of Modernity*; Kelly-Gadol, "Social Relations of the Sexes"). Household life in each of its global cultural settings—in its ethics, responsibilities, and priorities—has arguably the densest and most psychically compelling configurations of what modernity has defined itself against. Yet households are not going to disappear or wither away—ever. Moreover, the household and its kin relations are where the most stubborn resistance is found to imperial and colonial projects, as noted earlier. Of course, not all households are wonderful for women and children. Households are also powerful sites of violence against them and sites of their economic and political disempowerment. For these reasons, it seems a big mistake for progressive action groups to ignore women's lives in households as an origin of potential economic, political, and social insight and of progressive social transformation; to do so is to take sides, intentionally or not, with what is arguably the most resilient of patriarchal and imperial projects. The proposal here positions the women who are responsible for household life, in all its diversity and complexity, as the speakers of, the agents in, the ones whose interests and questions are also important for generating and evaluating the effects of scientific and technological research projects.

Second, such a project inserts the ethic of responsibility and care arising from household labor, as delineated by so many feminist political philosophers, into the desired ontology of research and political projects (Gilligan, *In a Different Voice*). Third, the temptation in "starting off from women's lives" has often been to think of women as autonomous individuals who are only contingently the center of households, kin relations, community relations, and relations with certain parts of natural environments. This temptation may arise when researchers think of "women's lives" primarily in terms of women in the West who work for wages in the formal economy and outside the household. This proposal reduces such residual excessive individualism, which damages so much of even progressive research and public policy.

Fourth, widespread progressive activist projects already do begin their research and activism from the standpoint of what happens in households. One example is the continuing projects around the globe which are focused on ending family violence—the child abuse which leaves such deadly residues in subsequent adult lives and the domestic battery and rape which so damage women and their children. Households and kin relations are not the only sites of such violence, of course; workplaces, the streets, and even churches are also not safe places for women or children. And then there are the horrors of war zones. Yet households do remain one site of opportunity for addressing violence against women and children. Such issues have remained crucial to advocacy of women's rights within national and international organizations and agencies. How do proposed scientific and technological transformations promise to increase or decrease women's vulnerability to such forms of domination and oppression? Other kinds of progressive projects where responsibility for women and households are central can be found in the older "wages for housework" movements. Yet others are the projects of women grassroots activists who have learned to function in international politics. No doubt there are others. Such projects show how to do research and activism that begins with what happens to women in households. Natural science projects, too, can begin with the costs to women in their household responsibilities as a result of supposedly progressive modernization projects in the realms of pharmaceutical, medical, agricultural, manufacturing, environmental, energy, and water improvement projects.

Finally (for now), even though modernity narratives and policy consistently position households in the sphere of tradition, today households breach so many of the binaries central to modernity's projects. Private vs. public life, pre-modern vs. modern, local vs. global or universal, reproductive vs. productive—today households around the globe do not easily fit into modernity's narratives except insofar as they remain primarily the responsibility of women.

Let us turn to consider some of the main objections to this proposal and so identify its additional strengths and limitations.

5. OBJECTIONS

We need not spend much time on the traditional male-supremacist view that households are "havens from the heartless world," where private life can be protected from encroachments by the harsh politics and economics of the public sphere. Of course feminists have had a great deal to say about this delusion. The one familiar but relevant point that I will mention here is that households are not only workplaces for producing things and services used by the household itself; they are also sites of production for exchange and sites for the organization of community political and social life more generally. As mentioned earlier, it was precisely the removal of such economic and political activities to the public sphere (as well as of education, care of the sick, and moral/religious education) that was a top priority for modernization policies. However, within supposedly modern societies, production for sale of agricultural products, food, clothing, indigenous pharmaceuticals, and other merchandise still occurs in households, as does domestic labor for other households, such as taking in washing and sewing, child care, and health care (e.g., midwifery). And the reproductive labor assigned to households is a necessary condition for the very existence of the productive world of work and politics. It is categorized as an "externality" in conventional economic theory, but it has calculable economic value. Such "traditional" household activities are not replaced during modernization, though, of course, it may be paid women workers who do much of it, either in households or in hospitals, nursery schools, or fast food enterprises. According to feminist economists, households are the site of one third to one half of human economic activity, including households in modern societies (cf. Sassen, *Globalization and Its Discontents*).

Furthermore, the Internet and cell phone have made possible many kinds of telecommuting and home work at all economic levels, from the artisanal work of dressmakers and caterers to craft manufacture for Internet sale, home-based financial and computer piecework and consultation, and the increasingly common office practice of working at home several days a week. These are just some of the contemporary practices which raise the interesting question of whether modern societies are becoming less modern as the purportedly required disaggregation of social institutions seems to be declining (a point raised earlier). My point is that households continue to play a much more important role even within supposedly modern societies than the modernization theorists recognize.

A second objection may come from feminists, who will protest that this proposed methodological move colludes with conservative tendencies to see women only as mothers or housewives (Mies, *Patriarchy and Accumulation*). It works against the huge effort feminists have made to get employers, governments, and agencies to see women as legitimate and valuable actors in public worlds also. To gain access to our professions and to continue to thrive within them, we have virtually had to deny that we even live in households that contain more than our completely autonomous selves and perhaps a completely self-sufficient partner! Certainly no sick children or partners or ailing parents, let alone pregnancies, would ever interfere with our devotion to a successful professional career, we have had to agree, or at least clearly imply. The struggle to get universities to provide time off for pregnancies or elder care, let alone sufficient child care, for faculty, students, and staff, is one kind of evidence of the legitimacy of women professionals' probable hostility to my suggestion here. Another kind is provided by the fact that even though wage discrimination against women in general in the United States, for example, has decreased over the last few decades, such discrimination against mothers remains virulent

(Correll, Benard, and Paik, "Getting a Job"; Crittenden, *Price of Motherhood*). To be sure, the disembedding of middle-class women from motherhood and households has been an important counter to the persistence of the public-sphere vs. private-sphere policies characteristic of dominant conceptions of social progress.[9] The proposal here certainly does not support the view that women are naturally, properly, or primarily housewives and mothers, but only that public policy, and science and technology policy and philosophy in particular, need to be "spoken" also by whoever is responsible for those kinds of crucial household activities and social relations. We can demand that public policy make the production of knowledge to serve the needs, interests, and desires of whoever is responsible for households as important as the flourishing of profiteering and militarism. Obviously, one solution is to get men—those in the privileged groups—to also take on such responsibilities. With wives and mothers working outside the household, many men already have. Yet their workplaces do not provide resources to them for taking on such responsibilities any more than they do to women, and men in those workplaces have rarely taken the lead in demanding such resources.[10]

Another objection is that women are in lots of places besides households, such as workplaces, community organizations, and national and international politics. So why focus only on their lives in households? Of course, starting to think about any issues from the standpoint of women's lives within those other sites is crucial for understanding how those sites actually work. Any such project should lead to recognition of the balancing act women must do to successfully juggle household and non-household responsibilities. Yet such projects all too often conceptualize women as fundamentally autonomous individuals, like their male coworkers, who just happen also to have additional responsibilities at home; this secondary feature is represented as handicapping them in the public sphere.[11] Looked at this way, public policy and practice, including science and technology policy and practice, obstruct the flourishing of the very households without which there would be no "autonomous individuals" to function in modern contexts. Why not think of women instead as responsible for reproducing human life itself, including the daily conditions for its flourishing, as well as often for the kinds of work formerly assigned only to men?

Yet another objection could be that not all women are in households or have such responsibilities at all. Women in all economic classes now work outside households and often live alone. In the new domestic and immigrant "serving classes" of support workers required by the new global information society, many women certainly are not living where their children, parents, and other kin live. They are nurses from the Philippines, nannies from Guatemala, and housekeepers from Mexico or North Africa who work in Los Angeles or Paris. Yet a closer look reveals that most women laboring far from home are probably still the centers of new forms of "distributed households"; their households are globally distributed. In agriculture, manufacturing, health care, and service industries in the First World can be found armies of men and women laborers, often immigrants, who produce the food consumed by elites, manufacture their clothes, and tend them in doctor's offices and hospitals. They are the millions of housekeepers and child-care workers needed to perform the traditional work for households where the women in those households work in the public sphere. Women have a high representation in these new classes of migratory international workers.[12] Women (and men) often leave their children and other dependents behind as they come to seek income at the global sites where the technically elite work is done. Moreover, among the technological elites in professional classes, such "distributed households" are also not hard to find. Many women working in professional jobs still have responsibilities, more than their brothers or husbands, for the needs of their adult children and their parents. So even women not living in their households often have major responsibility for the flourishing of their children, parents, extended families, and the communities in which they reside.

Now we are at a point to be able to see that new forms of families and households have emerged in the course of modernization's latest stage: global restructuring ("globalization").[13] It is families, households, and social communities that have also been globally restructured, not just the world of corporate profit-seeking and international relations. What we have in both the immigrant and new technical-elite classes are increasing numbers of distributed households, in which women from the South working in the North maintain far-flung households and kinship communities. They provide financial resources; they are actively involved in parenting through frequent telephone contact with their children, who are on their own or living with other kin; they engage in material and emotional care of parents and siblings, visiting as often as possible. Parallel kinds of household and kin relations are maintained by women in the technical-elite classes as their children, parents, and siblings become distributed through regional and international political economy networks.[14]

My point here is that by the households in which women live I intend to draw attention not only to the conventional bourgeois model of the nuclear family, against which feminists have long struggled and in which increasingly small proportions of women live, but rather to all of the motley and creative social arrangements women (and men) make to enable their dependents to survive and flourish.[15] These are excellent sites from which to ask questions about the conceptual and material practices of power advanced by particular science and technology projects.[16]

This leads to the final objection to be considered here. Scientists, engineers, and philosophers of science will have their own vigorous objections to this proposal. Certainly central to such resistance will be the question of the apparent absurdity of assuming that one should start off research from the lives of women in households—in the South as well as the North—to learn more about nuclear fission, whether the ozone hole can be closed, or just how it is that scientific practices come to constitute the very phenomena that they study. "*This* is a legitimate part of science?" Yet neither the science studies scholars of Part I nor the social movements of Part II conceptualize science as just lab or field studies isolated from their economic, political, and social preconditions, surrounds, and consequences. We can infer that they also would not conceptualize philosophy as simply working at the computer to channel pure thought onto the printed page, though little attention has been given to the social conditions of the production of philosophies of science. (However, see Mirowski, "How Positivism Made a Pact.") It was just such narrow conceptions of science and philosophy which represented scientists and philosophers as forbidden to evidence any interest in how the outcomes of their research or scholarship could be expected to perform or be deployed in daily life (with some exceptions, such as "mission-directed" health or environmental research). Once one conceptualizes all scientific and philosophic work as mission-directed to a greater or lesser degree and recognizes that we sometimes are not aware of the missions for which our funders or our cultural surround intends our work, then the proposal here will not seem quite so absurd.

The causal threads between abstract thought and the social worlds which make possible such thought, worlds on which such thought will have real material effects, seem especially long and fragile here. But so, too, did the links between the discoveries that made possible the pasteurization of milk and the social conditions that made possible the research, and on which that research had effects, before Bruno Latour (*Pasteurization of France*) excavated them. The proposal here will still be difficult to pursue, however, because we do not have available many models for how to think about how research on nuclear fission, the ozone hole, or the philosophy of science affects the flourishing of households and those responsible for them—to understate the issue! But this lack creates a splendid opportunity for new kinds of research agendas which can provide resources to advance both the growth of knowledge and social justice.

6. CONCLUSION

As long as the modern conceptual framework of philosophy, sciences, science studies, and other research disciplines in the West leads us to believe that what happens in households around the globe constitutes obstacles to the advance of objective and reliable knowledge projects and to the achievement of social justice, even progressive research and activism guided by such a framework is doomed to defeat. Women and modernity's Others cannot achieve social justice and make social progress on their own terms as long as this modernity vs. tradition contrast continues to shape the exceptionalist and triumphalist ways in which privileged groups (and the research disciplines that serve them) think and interact with others and the world around them.

Of course there cannot be only one solution to the multidimensional problems created by obsessive insistence on the contrast between modernity and tradition. The power of this contrast creates different problems for peoples differently affected by it. We have seen that Western modernity cannot escape tradition since its very existence depends upon tradition's persistence. My point is that it is better to choose which elements of one's past to retain and which to abandon than to remain haunted by specters of peoples and times one can no longer comprehend. Moreover, for those groups that modernity locates at its boundaries, it should be important to turn their problems with modernity and its representations of tradition into the problems of everyone.

What are the costs to progressive research and social transformation projects of ignoring the criticisms seen in earlier chapters of the contrasts between modernity and tradition, or of providing only partially adequate solutions which cannot comprehend how gender and racial/imperial discourses recirculate each other in the modern West? My provocation here is intended as just one possible way of raising such issues. Readers will no doubt come up with others.

Yet this moment is one of extraordinary opportunities for those in the West who are unhappy with this situation to join with those at the borders of conventional Western modernity in envisioning fresh ways to move forward in the production of reliable knowledge that can be *for* comprehensive social progress. The "crisis of the West" and the "crisis of masculinity," which seem to provide strong support for each other, can begin to be resolved with resources that already lie at hand. And Western scientific rationality and technical expertise can put down their heavy burdens of trying to deliver to the world forms of modernity which, it turns out, in large part block both self-understanding for the West and social progress for the multitudes.

NOTES

— — — — —

INTRODUCTION

1 "North/South" became the favored way to refer to the industrialized/
non-industrialized societies of the world more than a decade ago in
the context of the United Nations conference on environmental issues
held in Rio de Janeiro. This contrast replaced "First World/Third
World," which was rejected as an artifact of the Cold War, and "devel-
oped/underdeveloped," "haves/have nots," "West/Orient," and earlier shorthand ways of referring to the effects of five centuries of
European and North American imperialism, colonialism, capitalist
expansion, and the diverse other local politics which have bequeathed
us contemporary global social relations. No such contrast is entirely
accurate and most carry regressive political meanings. Moreover, any
such contrast inaccurately homogenizes the two groups and obscures
more complex social relations between and among various global
groupings. In this last category are the changing roles of the "Second
World" (the former Soviet bloc) and the emerging "Fourth World"
(Third World communities in the urban centers of the First World).
Additionally, progressive groups in the Third World have sometimes
adopted these now-rejected names for their own projects. Another
problem is that such contrasts reify a preoccupation with differences
that hides shared interests between peoples in very different social
circumstances. Yet it would be premature to avoid all such binaries
and thereby make invisible global patterns which create radically dif-

ferent life conditions for people who happen to be born into one society rather than another, or into one family rather than another in any given society. Similarly, one can't make male supremacy go away by refusing to indicate which are the women and which the men, or dissolve exploitative class relations by refusing to recognize which people are poor and which are rich. Here I shall favor "North/South" in spite of its geographical inadequacy and the other problems indicated, but use the earlier distinctions when referring to the eras in which these other terms were commonly used.

2 See, for example, Gross and Levitt, *Higher Superstition*; Gross, Levitt, and Lewis, *The Flight From Science and Reason*; Kimball, *Tenured Radicals*; Sokal and Bricmont, *Fashionable Nonsense*; Weinberg, "Revolution that Didn't Happen." Cf. Ross, *Science Wars*.

3 See, for example, Hecht and Anderson, "Postcolonial Technoscience"; Hess, *Science and Technology*; and Selin, *Encyclopedia of the History of Science*. Cf. Harding, *Is Science Multicultural?*

4 Sharon Traweek helped me to understand my project in this way.

5 This followed the issue on "early modernities" around the globe, not just in Europe (Eisenstadt and Schluchter), to which we return in a later chapter.

6 What is at stake in these different periodizations of the modern? Especially what is at stake for those that center on the Scientific Revolution and the English Revolution vs. those on the Enlightenment or on the French Revolution and the Industrial Revolution? (David Hess has raised this question.) I suspect that both disciplinary and broader political interests create the intense commitments one can find to such distinctive foci, but must pass this question on to historians and sociologists of knowledge.

7 Doug Kellner makes this point in an unpublished manuscript.

8 As we will see in Part III, they may be post-premodern—that is, modern—but that does not mean they are post-traditional. The premodern and the traditional are not identical.

9 For example, gender will be discussed in chapter 3.

10 I thank Jim Maffie for making me think about what my investments are in the term "empirical" in discussions such as this one (" 'To Think with a Good Heart' "). Maffie is a scholar of Native American philosophy before the European conquest. He argues that since many of the spiritual ontological assumptions and methodological practices of the Nahua contribute to their production of reliable knowledge of the world around them, and no less than have Western religious and cultural beliefs advanced modern Western sciences, my emphasis on the term "empirical" still restricts what counts as science in a Euro-

centric manner. He may well be right, in my considered opinion. Indeed, Maffie's point could get support from such recent Northern science studies accounts as that of Gibbons, Nowotny, and Scott of the inevitable and desirable social "contextualization" of the production of scientific knowledge (see chapter 3). Maffie's issue deserves further discussion. Nevertheless, I will continue to use the term in such discussions since I also face the challenge of making my arguments plausible to readers who will be encountering postcolonial science studies—and perhaps even post-Kuhnian Northern science and technology studies—for the first time and will still hold exceptionalist beliefs about modern Western sciences. For them I must make clear that non-Western sciences provide generally reliable guides to the interactions of non-Western cultures with natural and social environments.

11 And, like Beck's, they lie outside the subfield of the new sociology of scientific knowledge, whose valuable microstudies of the constitution of nature and truths about it in scientific labs and field sites—of which Latour was a pioneer—have become institutionalized as the most "scientific" of the "sciences of science" which form science and technology studies. As is indicated by my choice of Beck and GNS to also represent such sciences of science, I think this too narrow a conception of the necessary tasks of science studies.

1. MODERNITY'S MISLEADING DREAM

1 Of course there are a number of brilliant scholars of nationally Northern origin—from the United States, Canada, Australia, Scandinavian countries, etc.—who have helped to create the field of postcolonial science studies. There are also numerous scientists and science studies scholars living in the South who are enthusiastic defenders of mainstream Northern science studies as well as of positivist approaches. "Northern science studies" here refers to a theoretical and political orientation of mainstream science studies which is importantly postpositivist yet still functionally Eurocentric (and often androcentric), whatever its intent; the phrase does not refer to postal addresses or ethnicity per se. It should be pointed out that interest in looking past this kind of modernity horizon is beginning to emerge in an oblique way through anti-Eurocentric studies of indigenous knowledge and traditional environmental knowledge, and especially in a still relatively small number of comparative studies of Northern and indigenous or traditional knowledge systems. I pursue these issues further in later chapters.

2   Latour's rhetorical style is so engaging, and it so plausibly conveys what can otherwise seem to be improbable proposals, that I provide in this chapter several extended passages in his own voice.

3   "Postmodernism is a symptom, not a fresh solution. . . . It senses that something has gone awry in the modern critique, but it is not able to do anything but prolong that critique, though without believing in its foundations. Instead of moving on to empirical studies of the networks that give meaning to the work of purification it denounces, postmodernism rejects all empirical work as illusory and deceptively scientistic. Disappointed rationalists, its adepts indeed sense that modernism is done for. . . ." (*We Have Never Been Modern* 46).

4   The Science Wars mobilized scientists against the field of science studies, and have often specifically targeted Latour's earlier work (as well as the work of Donna Haraway and the present author). See Gross and Levitt, *Higher Superstition*; Gross, Levitt, and Lewis, *Flight from Science and Reason*; Kimball, *Tenured Radicals*; Ross, *Science Wars*; and Sokal and Bricmont, *Fashionable Nonsense*.

5   I return in the third section of this chapter to Latour's inadequate engagement with postcolonial science studies. With respect to feminist science studies, in the more recent of the two books considered here, we find a sentence on p. 33 which praises how "feminists have shown often enough how the assimilation of women to nature had the effect of depriving women of all political rights for a very long time." Then, in a few lines on p. 49 and in two footnotes (p. 260, notes 57 and 58), he praises the "immense work of the feminists" in showing the problems with marking only women with gender. Here he recommends the "enormous analytic work" summarized in Evelyn Fox Keller's *Reflections on Gender and Science*; Schiebinger's *The Mind Has No Sex?*, which links feminist questions and science studies; Donna Haraway's feminist political ecology (*Primate Visions*; *Simians, Cyborgs, and Women*; *Modest Witness*); and feminist work in sociobiology in Strum and Fedigan (*Primate Encounters*). But we do not see how this work has any effect on Latour's account.

6   Plato is certainly a culprit here, too. An ancient form of this practice can be found in Plato's division of humans into those with souls of gold, silver, or lead, and thus by birth suitable for lives as rulers, in militaries, or as artisans respectively.

7   Readers unfamiliar with political philosophy need to be reminded that the term "republican" as used here does not refer to the current platform of the U.S. political party by that name but rather to a particular early form of democratic governance which can trace its inheritance to the French Revolution and to Rome.

2. THE FIRST MODERNITY OF INDUSTRIAL SOCIETY

1   In this chapter citations to major sources in Beck's writing will be referred to by the initials indicated in each case: *Reinvention of Politics* (RP); *Risk Society* (RS); and *World Risk Society* (WRS).

2   The phrase "science of science" has a history in Marxian thought. However, we need not pursue that issue here since both Latour and Beck clearly use it to refer to the new, post-Kuhnian sociologies (and ethnographies and histories) of science.

3   Here I am trying to make sense of what I take to be Beck's use of the term "reflexivity" to refer to two distinct phenomena at different points in his writings.

4   Here and above are typical examples of how Beck admirably recognizes the power of feminist analyses and yet does not engage them. What is the content of these analyses? What are the nonprofessional "rationality and praxis" and "specialist competence" which appear in such analyses? How have these feminist projects illuminated or changed professions and occupations? How have they informed his own account?

5   This is the notion I developed into "Robust Reflexivity" in *Is Science Multicultural?* See also Elam and Juhlin, "When Harry Met Sandra."

6   Sometimes Beck insists that he always means both of these two definitions of "reflexivity" and in other places that he only means this second notion. For an example of this second position, see his reply to Lash's essay ("The Reinvention of Politics" in Beck, Giddens, and Lash, *Reflexive Modernization*), where he is annoyed that Lash attributes to Beck the self-critical meaning of reflexivity—the sciences of science—rather than the unpredictable consequences of modernity and its sciences, on which Beck wants to focus.

7   See also Castells, *Power of Identity*, on this phenomenon.

8   Beck's account here echoes central themes in the multiple modernities discussions, including the interrogation of the category of tradition. These topics are pursued in chapters 6 and 7.

9   For example, see New ("Class Society or Risk Society?"), Draper ("Risk, Society, and Social Theory"), Hall ("Review of *Risk Society*"), and Boyd ("Review of *Risk Society*") on *Risk Society*; and see Robertson ("Review of *The Reinvention of Politics*"), Linklater ("Review of *The Reinvention of Politics*"), Mike Smith et al. ("Reinvention of Politics"), and Boyne ("Politics of Risk Society") on *Reinvention of Politics*. But see also Adam, Beck, and Van Loon, *Risk Society and Beyond*.

10 This insight begins to grasp some of the central assumptions of the multiple modernities discussions to be considered in chapter 6.

11   Feminist critics of modernization theory will spell out in a different way some of the consequences of such "achievements."

12   See, for example, Haraway, "Science Question in Feminism" and my work to which Haraway is here responding (*Science Question in Feminism*), as well as my later work, e.g., "Instability of the Analytical Categories."

13   For an earlier discussion of the importance of such movements, see Rose and Nowotny, *Counter-Movements in the Sciences.*

3. CO-EVOLVING SCIENCE AND SOCIETY

1   They insist that the three of them are equal contributors to their joint work and that the order in which their names appear as authors is purely arbitrary (*Re-Thinking Science*, viii). In the first book examined here Gibbons is first author, and in the second one Nowotny holds this place. I shall cite issues raised specifically in one of the books by using the initials of the title (NPK for *New Production of Knowledge* and RTS for *Re-Thinking Science*), and to themes and claims that appear in both by referring to GNS.

2   Other observers subsequently also have pursued the value of thinking about the topic of the co-evolution of science and society (Jasanoff, *States of Knowledge; Designs on Nature*). In retrospect, the language of evolution seems to have been destined to become widespread as a way familiar and, presumedly, acceptable to scientists with which to capture the ongoing relations between science and society. Use of the evolution metaphor tends to naturalize the processes so described and thus to depoliticize them. It makes active political engagement seem irrelevant. Who can presume to change the direction of human evolution? My point here is not about the intentions of these authors but about the implications of their conceptual framework, intended or not.

3   Chapter 6 in RTS also pursues this issue.

4   I shall rely more on the ways in which they present their arguments in the later book since that one has had the benefit of more extensive reflection and observation.

5   The new historians of science would say that it had always been more a figment of science's political rhetoric than an actual practice. See, e.g., Forman, "Behind Quantum Electronics"; Rose and Rose, "Incorporation of Science." I return to this point below.

6   The language here is Karl Popper's in *Conjectures and Refutations.*

7   See here also a valuable criticism of the defense of a "hard core" of cumulative approaches to scientific truths vs. the "soft" images of it

which do change from era to era and culture to culture (RTS 188ff.). Such influential defenses can be found in the work of the physicist Steven Weinberg ("Revolution That Didn't Happen") and the philosopher Yehuda Elkana ("A Programmatic Attempt"), for example.

8   Recollect that Beck, too, was criticized for this kind of error.

9   See Mirowski, "How Positivism Made a Pact," for a discussion of the failure in contemporary science studies to fully investigate the "social" in "the social construction of knowledge"—an issue to which I return in subsequent chapters.

4. WOMEN AS SUBJECTS OF HISTORY

1   I thank David Hess for stimulating me to clarify this decision in this way.

2   Of course the field of science studies has itself been a social movement very much positioned against dominant disciplinary understandings and, in some cases, practices of the production of scientific knowledge. So all three contributors discussed in Part I do share a kind of solidarity against the political and scientific elites which have produced the kinds of sciences and politics that serve their interests. My point in this book is that such solidarity could and should be shared between those mainstream science studies figures and the feminist and postcolonial scholars who are discussed here, yet all sides seem unclear about what is to be gained by such alliances. So in this section it is three neglected, or even banished from the mainstream, fields of science and technology studies which are our topic.

3   For philosophers, this new field began with the work of Willard Van Orman Quine, who in 1953 already had challenged "Two Dogmas of Empiricism." It is significant that several of the early feminist philosophers of science (including this author) did their dissertations on Quine. Kuhn's *Structure of Scientific Revolutions* then stirred up a flurry of largely negative response from philosophers (see, e.g., Lakatos and Musgrave, *Criticism and the Growth of Knowledge*). Other important early works in the Northern movement included Hagstrom, *The Scientific Community*; Machlup, *Production and Distribution of Knowledge*; Price, *Little Science, Big Science*; and Ravetz, *Scientific Knowledge and Its Social Problems*. Cf. discussions of this era in Hollinger, *Science, Jews, and Secular Culture*.

4   For surveys and conceptualizations of the early work see Fausto-Sterling, *Myths of Gender*; Harding, *Science Question in Feminism*; Harding and O'Barr, *Sex and Scientific Inquiry*; Hubbard, *Politics of*

Women's Biology; Keller, *Reflections on Gender and Science*; Schiebinger, *The Mind Has No Sex?* See also Donna Haraway's work, and Wajcman, *Feminism Confronts Technology*. See Potter, *Gender and Boyle's Law*, for an unusual feminist foray into a field imagined to be immune to such analysis.

5 Though even in the philosophy of science, which of the science studies fields is still the most constrained by positivist commitments, this picture is beginning to change. See Figueroa and Harding, *Science and Other Cultures*, which is one of the dissemination vehicles for a grant from the National Science Foundation to the American Philosophical Association to organize lectures and publications on diversity issues in the philosophy of science.

6 Of course women were present in many other jobs necessary for the production of scientific knowledge—as teachers, textbook authors, illustrators, data collectors and analyzers, lab technicians, secretaries, popular writers, and, of course, in cleaning, food preparation, and all the other services to the public institutions and domestic worlds necessary to keep scientists alive and functioning. Women were just not in the peer group of scientists.

7 As some women have joked, "testosterone poisoning" in his own case, and possibly that of other male Harvard professors, evidently was not one of the biological differences on his mind!

8 For example, in societies with high professional sex-segregation, such as Islamic or some Catholic societies, where only women doctors were permitted to treat women, only women teachers to teach girls, women lawyers to represent their women clients, and so forth, women have tended to hold higher-level professional positions than in more sex-integrated societies. To take a different kind of case, in recently developing Third World countries, such as India, the government made a huge investment in technology research and training. Boys but not girls were targeted for this training in the early years. Tertiary technology education was located mostly in new technology institutes, leaving physics behind in the older universities, where women were more welcome than in the new technology institutes. And there are other local histories and practices which have contributed to higher representation of women in the sciences, mathematics, and engineering than in Western Europe and North America. See Koblitz, "Challenges in Interpreting Data" for a discussion of this phenomenon.

9 Of course this was a main focus of the three science studies scholars discussed in Part I.

10 See Jane Flax's influential discussion of what gender is in her "Gender as a Social Problem."

11 See Fausto-Sterling, *Myths of Gender*, and also *Sexing the Body*.

12 These days there is a startling increase in women heads of state, from Angela Merkle in Germany to the president of Chile. The head of the Pan African Union is a woman. Does this signify the successful mainstreaming of feminist assumptions and demands? No doubt. The already or soon-to-be decreasing power of states and state-like governments? Someone else will have to assess this possibility.

13 This history can be found in Hartsock, "Feminist Standpoint"; Jaggar, "Feminist Politics and Epistemology"; Jameson, " 'History and Class Consciousness' "; Pels, "Strange Standpoints."

14 I have had a hand in developing and disseminating standpoint approaches ever since Harding and Hintikka, *Discovering Reality*. See, most recently, Harding, *Feminist Standpoint Theory Reader*.

15 As indicated earlier, such theories were developed within a number of different disciplines with diverse histories and preoccupations, and by theorists with commitments of varying strength to Marxian and to Enlightenment projects. Consequently, it is risky to try to summarize this approach in any way that attributes to it a unified set of claims. Nevertheless, theorists from these different disciplines do share important assumptions and projects that differ from conventional understandings of what makes good science, including, I propose, the features identified in this section. (I articulate them in a form which stays close to Hartsock's account: "Feminist Standpoint" 1983.) Of course not every theorist equally prioritizes or emphasizes each of these, since what is perceived to be important in the context of sociology may be less important to political philosophers or philosophers of science and vice versa. Nor are disciplinary concerns, themselves heterogeneous, the only ones that lead to divergence in how standpoint approaches have been developed. The particular research projects of standpoint theorists have also shaped how they use a standpoint logic.

16 Note that this theme echoes standard beliefs about the effectiveness of scientific methods: which interactions with, or kinds of observations of, natural and social worlds are pursued both enables and limits what one can know.

17 I use the term "ideology" here to mean systems of false interested beliefs, not just of any interested beliefs.

18 Donna Haraway famously developed this notion of socially situated knowledge in a paper which originated as a comment on my *Science Question in Feminism*. See her "Situated Knowledges."

19 A motto from the early days of the women's movements of the 1970s was "The degree of his resistance is the measure of your oppression." If this point is lost, and even some standpoint defenders sometimes

lose it, "standpoint" seems like just another term for a perspective or viewpoint. Yet the standpoint claim about the epistemic value of some kinds of political struggle—the epistemic value of the engagement of the researcher—is thereby made obscure when its technical use, which I retain here, is abandoned.

20   I shall refer to standpoint approaches as inherently progressive since that is the way they have been understood today through the Marxian legacy inherited by leading movements for social justice. Yet it is useful to recall that Nazi ideology also (ambivalently) opposed modern science on standpoint grounds and, indeed, conceptualized its murderous program as one of advancing social justice (see Pels, "Strange Standpoints"; Proctor, *Racial Hygiene*). Religious fundamentalist, geographically based ethnic, and patriot or neo-Nazi social movements usually are not reasonably characterized as dominant groups. Nevertheless, they too are threatened by the political values and interests of modernity which sciences represent. They often make something close to politically regressive standpoint arguments. So theories about which kinds of social movements are liberatory, and for whom, must be articulated to justify research projects in the natural and social sciences. See Castells, *Politics of Identity* for an interesting discussion of the different political potentialities of various identity-based social movements around the world today.

Of course there is nothing new about natural and social science research *assuming* political theories. Conventional philosophies of natural and social science always assumed—consciously or not—Liberal political philosophies and their understandings of relations between knowledge, politics, and social emancipation. Indeed some observers have argued that a coherent philosophy of science and of politics was exactly the goal of early modern theorists (see, e.g., Shapin and Shaffer, *Leviathan*). Sciences and their philosophies are always at least partially integrated into their larger economic, political, social formation, to put Kuhn's point another way (cf., e.g., Schuster and Yeo, *Politics and Rhetoric*; Steinmetz, *Politics of Method*). Thus it is not standpoint theory that introduces the conjunction of social theory (or political philosophy) and philosophies of science or knowledge, let alone their "integrity" with actual historical features of a society.

21   The term "subject of social science" can be confusing to social scientists for whom their "subjects" are what the rest of us might refer to as the objects of their studies. From the perspective here, it is the social scientists collectively, and perhaps their disciplines, funders, and sponsors, who are the subjects—the speakers, the "voice"—of those studies.

22   See Fausto-Sterling, *Myths of Gender* and "Refashioning Race" (forthcoming); all of Haraway's work; Schiebinger, *The Mind Has No Sex?* and *Plants and Empire*; Schiebinger and Swan, *Colonial Botany*, for the work of influential Northern feminist science studies scholars which has in practice as well as in principle brilliantly focused on women's different conditions around the globe and as such has made important contributions also to the postcolonial science studies work to be considered in the next chapter.

23   Rouse ("Feminism and the Social Construction of Knowledge") in an otherwise illuminating essay, argues that the sociology of scientific knowledge also is politically engaged. His evidence is a few statements by several such authors claiming that their work supports a renewed "humanism." They have in mind that the nature we engage with is always already entered into human social relations; it cannot stand as a socially neutral resource for deciding conflicts within social relations. This was Latour's point also. For the feminist scholars, this "social constructivist" stance, which they share, doesn't yet address the gender issues toward which their political engagement is directed. It does not center social and political inequality.

24   "Liberal," in this book, is capitalized to refer to the social contract theory grounding democratic revolutions of the eighteenth century and the new states they created.

25   Though Latour would include a few individual feminist scientists and science studies figures as inside the sciences of science, he would not include the whole field since, to him, it would represent a reprehensible example of "identity politics."

26   One important exception to this claim is to be found in feminist environmentalism. See, e.g., Seager, *Earth Follies* and "Rachel Carson Died of Breast Cancer."

27   See, for example, the work cited above in note 22.

5. POSTCOLONIAL SCIENCE AND TECHNOLOGY STUDIES

1   In the introductory chapter I explained how embedded in various eras of Eurocentric politics were all of the conventional ways of referring to the macrosocial structures which have shaped global social relations from 1492 to the present day: the West vs. "the Rest" or "the Orient," First vs. Third World, underdeveloped vs. developed. None represent the emancipatory and pro-democratic ways in which Europe's "Others" think of themselves. Arguably the politically most neutral language is the North vs. South contrast which emerged from the 1992

UN Earth Summit in Rio de Janeiro. Yet it is geographically confusing: Are Japan and South Korea in the North or the South? And what about the former Second World of countries that were aligned with the Soviet Union during the Cold War? Does it make sense to conceptualize them as in the North or the South? What about the huge presence of non-Western immigrants in the major industrial and trade centers of the North—New York, Los Angeles, London, Amsterdam—and the redistribution by globalization and warfare of Third World workers and refugees over the world, North and South? As I indicated in the introduction, I shall use whichever language best articulates the politics of the era referred to. I hope that this practice is not confusing reading in this chapter in particular, where my concern is past and present global social relations between "the West" and those societies it "others."

2   See, for example, a presidential address to the Anthropology Society of Washington in the late 1950s by Frake, "Ethnographic Study of Cognitive Systems."

3   All of these four projects have come to be referred to as part of PCSTS, even though China and Japan, for example, were not European colonies, and Latin American states mostly gained formal independence in the nineteenth century. Thus the "post" of "postcolonialism" refers to different eras for India and Africa than for Latin American societies, and is technically not appropriate for China and Japan. To some extent the situations of Africa and India (each itself internally diverse) seem to have become the models for postcolonial science and technology studies, perhaps due in part to the widespread availability of highly educated English and French speakers and writers in many of these societies, thanks to the educational practices of the British and French empires. Yet this phenomenon can distort issues about relations between European and other scientific and technological traditions. Moreover, in light of neocolonial Western economic, political, and cultural practices, "postcolonializing" would be a better term than "postcolonial" for present-day accounts and practices. Yet the term is valuable as a way to name a discursive space within which new kinds of questions and issues can be raised. See Ashcroft, Griffiths, and Tiffin, *Postcolonial Studies Reader*; and Williams and Chrisman, *Colonial Discourse*, for more discussions of problems with the concept of postcolonial.

   Of course scholars of both Northern and Southern descent have actively and sometimes conjointly produced this work. And traditional Northern "international science" is widely desired and practiced outside Europe and North America. Indeed, it is not hard to find strongly critical responses from Third World intellectuals toward the reevaluation of indigenous knowledge traditions around the globe (cf. Nanda, *Prophets Facing Backward*, and responses to this book in Maffie, *Social Epistemology*; also Aikenhead, *Multicultural Sciences*).

The "North vs. South" framework here does keep in focus the history and present practices of the global political economy, but it also distorts the complex histories and evaluations of both Northern sciences and of indigenous knowledge systems, elements of which one can find in every society around the globe.

4   A few of the influential works in this field, in addition to those mentioned above, are Adas, *Machines as the Measure of Man*; Blaut, *Colonizer's Model of the World*; Braidotti et al., *Women, the Environment, and Sustainable Development*; Brockway, *Science and Colonial Expansion*; Crosby, *Columbian Exchange* and *Ecological Imperialism*; Goonatilake, "Voyages of Discovery"; Haraway, *Primate Visions*; Headrick, *Tools of Empire*; Hecht and Anderson, *Postcolonial Technoscience*; Hess, *Science and Technology*; Joseph, *Crest of the Peacock*; Kochhar, "Science in British India"; Lach, *Asia in the Making of Europe*; Maffie, *Truth*; McClellan, *Colonialism and Science*; Nader, *Naked Science*; Nandy, *Science, Hegemony, and Violence*; Petitjean et al., *Science and Empires*; Prakash, *Another Reason*; Pyenson, *Cultural Imperialism*; Reingold and Rothenberg, *Scientific Colonialism*; Selin, *Encyclopedia of the History of Science*; Shiva, *Staying Alive*; Third World Network, *Modern Science in Crisis*; Turnbull, *Masons, Tricksters, and Cartographers*; Verran, *Science and an African Logic*; and Watson-Verran and Turnbull, "Science and Other Indigenous Knowledge Systems." See also Harding, *"Racial" Economy of Science*, *Is Science Multicultural?*, and *Science and Social Inequality*; and Figueroa and Harding, *Science and Other Cultures*.

5   An eminent science studies scholar made such a charge at a History of Science conference in response to my work on PCSTS a few years ago.

6   These were identified in the discussion of feminist standpoints in the preceding chapter.

7   Some of the PCSTS accounts are overtly articulated within these post-Marxian frameworks. See, e.g., Brockway, *Science and Colonial Expansion*; and Sachs, *Development Dictionary*. Others appreciate the strengths of such projects and move beyond them to less problematic conceptual schemes.

8   Such as those we saw in the writings of the transformative science and technology studies scholars in Part I.

9   See also Blaut, *Colonizer's Model of the World*; C. L. R. James, *Black Jacobins*.

10 I first encountered these themes in the Caribbean economist Walter

Rodney's powerful account of how Europe underdeveloped Africa, as he put the point (*How Europe Underdeveloped Africa*). By a year or so later, a colleague at Howard University (a distinguished historically African American university in Washington) reported that this book had become a focus of student intellectual and political excitement; there were Rodney reading groups springing up on the Howard campus.

11   Cf. Adas, *Machines as the Measure of Man*; Brockway, *Science and Colonial Expansion*; Headrick, *Tools of Empire*; Hess, *Science and Technology*; Khor, "Science and Development"; Kumar, "Problems in Science Administration"; McClellan, *Colonialism and Science*; Nandy, *Science, Hegemony, and Violence*; Philip, *Civilising Natures*; Reingold and Rothenberg, *Scientific Colonialism*; Sardar, *The Revenge of Athena*; Schiebinger, *Plants and Empire*; Schiebinger and Swan, *Colonial Botany*; Third World Network, *Modern Science in Crisis*; Weatherford, *Indian Givers*.

12   Cited in Nader, *Naked Science*.

13   Two sources give a good introduction to the richness of this approach; see Selin, *Encyclopedia of the History of Science*; and the *Indigenous Knowledge and Development Monitor*. See also Hess, *Science and Technology*; Nader, *Naked Science*.

14   For a more extended discussion, see chapter 4 of Harding, *Is Science Multicultural?*

15   See Frake, "Ethnographic Study of Cognitive Systems."

16   Of course this account will be controversial. No doubt some of its claims will have to be adjusted to the perceptions of other scholars. It will be interesting to see how its claims fare in the next decade as postcolonial perspectives become more familiar to a new generation of scholars.

17   For more extended discussion of these possibilities, see Harding, *Science and Social Inequality* 54–61.

18   That is, including a modest form of delinking that would encourage science and technology creativity within each Third World culture.

19   And in some contexts a disadvantage if, for example, it interrupts the possibility of a necessary completion of a particular therapy.

20   Some of the influential such writings are Gross and Levitt, *Higher Superstition*; Gross, Levitt, and Lewis, *Flight from Science and Reason*; Kimball, *Tenured Radicals*; Ross, *Science Wars*; Sokol and Bricmont, *Fashionable Nonsense*. A recent echo of these debates can be found in Nanda, *Prophets Facing Backward*, with responses to this book in Maffie, *Social Epistemology*. In the interests of "full disclosure," my work has frequently been a target of these attacks.

6. WOMEN ON MODERNITY'S HORIZONS

1 See, for example, Lionnet and Shih, *Minority Transnationalisms*. I thank Françoise Lionnet for emphasizing this point to me (in conversation).

2 The West/Rest binary was discussed in the introduction and the gender binary in chapter 4.

3 See, for example, how Collins (*Black Feminist Thought*) wrestles with such issues. For an illuminating collection of reflections by distinguished feminist anthropologists, sociologists, and historians on how they problematically positioned themselves in some of their studies, see Wolf (*Feminist Dilemmas in Fieldwork*). See also Harding and Norberg, *New Feminist Approaches to Social Science Methodologies*. By now, useful collections of essays focusing on these and other methodological issues have appeared in virtually every social science discipline. Feminist work was not the first to identify these problems. Yet it has come to be in the forefront of social science work more generally in producing these kinds of usually anguished reflections on the relation of the observer to the observed, ever since the positivist dream of reflecting in our minds a world which is out there for the reflecting was revealed as a delusion. Rorty (*Philosophy and the Mirror of Nature*) is a classical location for this postpositivist perception. See also Steinmetz (*Politics of Method*) on postpositivist methodological issues in the social sciences.

4 This literature is dispersed in many disciplines and regional case studies, and in governmental reports. It can also be found in bits and pieces in critical feminist work on development, international relations, and globalization. An excellent source for the arguments, policy considerations, and significant literatures is the set of review essays commissioned by the United Nations Commission on Science and Technology for Development, produced by their Gender Working Group (*Missing Links*). With respect to the issues in this chapter, see especially the essays by Kettel ("Key Paths"); Appleton et al. ("Claiming and Using Indigenous Knowledge"); Wakhungu and Cecelski ("A Crisis in Power"); Yoon ("Looking at Health"); and Kazanjian ("Doing the Right Thing"). See also Harding and McGregor ("Gender Dimension"); Mies ("Patriarchy and Accumulation"); Shiva (*Staying Alive, Monocultures of the Mind, Close to Home, Stolen Harvest*, and her other books); and Visvanathan et al. (*Women, Gender and Development Reader*).

5 "Pre-modern" and "traditional," along with "the modern," "modernization," and "postmodern," will be problematized in the next two chapters. My citations in this section are disproportionately to older, highly

influential writings in order to draw attention to the familiarity in other contexts of the claims here.

6   Sara Ruddick's *Maternal Thinking* provided a powerful account of how any person who mothers children must think in distinctive ways about such work.

7   Slave infanticide in the U.S. South, done for the same reason, was the theme of the internationally famous novel *Beloved,* by the Nobel Prize–winner Toni Morrison.

8   Some readers might doubt that there are women's single-sex communities in the North. There are many. Examples are religious orders and activities within churches; colleges and sororities; occupations such as secretarial work, nursing, and elementary school teaching; and women's service institutions and organizations such as rape crisis and spousal-abuse centers.

9   See also the immense literature on gender and modernization cited in chapter 8.

10   In chapter 3 we saw the team of Gibbons, Nowotny, and Scott criticize this unrealistic conceptual divide between pure and applied research in today's research environment. We saw Latour and Beck criticize the view that a clear and useful divide between science and politics in fact could ever be found.

11   It is now widely recognized that it is poverty which increases population growth, not the reverse, as decades of international population policy theories and policies proclaimed. Poor people tend to engage in labor-intensive economic activity. They must depend upon children and kin for the health care, child care, elder care, and unemployment support which welfare systems provide to the middle classes in the West and wealth provides to the already most-advantaged everywhere. Poor people cannot afford to have small families. It turns out that educating women so that they have their own sources of cash income is the single factor most effective in reducing population size.

7. MULTIPLE MODERNITIES

1   Of course there are multiple scientific disciplinary traditions and practices within Western science—physics, chemistry, biology, geology, etc. The controversial issue here is whether there are multiple *culturally* distinctive knowledge-seeking traditions and practices that are as deserving as are Western traditions and practices of the term "science." Recollect that my point was not fundamentally about terminology but, rather, about interrogating the conventional distinctions

between modernity and tradition, value-free science and tradition-embedded "indigenous knowledge." Triumphalism and exceptionalism are the issues.

2   See, for example, the writings of Donna Haraway, James Maffie, Laura Nader, David Turnbull, Helen Verran, and those from the North (and, of course, the South) contributing to the monumental encyclopedia edited by Helaine Selin (*Encyclopedia of the History of Science*). Forerunners of this work can be found in some of the comparative studies familiar to Northern historians of science in which the strengths of non-Western empirical knowledge systems were emphasized. Cf., e.g., Needham, *Grand Titration* and *Science and Civilisation in China*; Sabra, "Scientific Enterprise"; and the many authors cited in Hobson, *Eastern Origins of Western Civilisation*. See also the citations in Harding, *Is Science Multicultural?*; and Hess, *Science and Technology*.

3   For discussion of this history see, e.g., Eisenstadt ("Multiple Modernities") and many of the other essays in Eisenstadt (*Multiple Modernities*); see also Heilbron et al., *Rise of the Social Sciences*.

4   The phrase is Ortiz's ("From Incomplete Modernity to World Modernity").

5   Does this history illuminate the resistance to postmodernism in the social sciences?

6   As these are commonly identified in Anglophone accounts. Francophone and perhaps other discourses sort these out somewhat differently (Friedman, "Definitional Excursions").

7   See, for example, Cockburn, *Machinery of Dominance*; Feenberg, *Alternative Modernity*; *Transforming Technology*; Ihde, *Technology and the Lifeworld*; MacKenzie and Wacjman, *Social Shaping of Technology*; Murata, "Creativity of Technology"; Wajcman, *Technofeminism*.

8   Murata cites this example of the "negative creativity" of technology, drawing on Tenner, *Why Things Bite Back*. It turns out that our desktops, with their more or less careful piles of paper, represent a valuable kind of map of our minds—the things we are working on or still thinking about—which computer menus don't (at least, don't yet) accurately represent.

9   See Cockburn and Ormrod, *Gender and Technology*; and Kleinman and Vallas, "Science, Capitalism, and the Rise of the 'Knowledge Worker,'" respectively, for the last two shifts.

10   Sharon Traweek's comparative work on high-energy physics in Japan and the United States is also revealing in this respect. Traditional Japanese work patterns and legal restrictions on modifying scientific equipment consort to make high-energy physics experiments in Japan tend to run longer than they typically do in Europe and the United

States (*Beamtimes*). Does this have consequences for what the Japanese and Westerners can know about nature's order? See also Galison, *How Experiments End*; and Galison and Stump, *Disunity of Science*.

8. HAUNTED MODERNITIES

1   Hereafter, Kelly-Gadol's "Social Relations of the Sexes" will be cited parenthetically as s r s.

2   Hereafter, Scott's *Gender and Development* will be cited parenthetically as G D.

3   Kelly-Gadol's use of the terms "sex" and "sex roles" where today one would speak of gender, or at least sex/gender, marks the radical character of feminist accounts of gender, men's no less than women's, as socially constituted. These accounts were just beginning to appear as Kelly-Gadol wrote. Today one also can find accounts of sex, and sex differences—not just gender and gender-differences—as socially constituted, which is possibly an equally radical understanding. See, e.g., Fausto-Sterling, *Sexing the Body*.

4   Interestingly, Beck and Beck-Gersheim (*Normal Chaos of Love*) do address the loss of status and power for women when work and family life separate in the context of the recent women's movements. They are especially interested in how what appear to be the "private troubles" of relationships these days—the "normal chaos of love"—in fact are simply reflections in private life of public troubles between the changing institutions of modernity. Now if only Beck and Beck-Gersheim had figured out how to connect these illuminating insights with Beck's critical rethinking of science and modernity! But that is an Amazonian task for anyone undertaking it, as we shall see.

5   Such knowledge is not normally recognized as "indigenous knowledge" because no women (and men) of European descent who are living in modern societies are recognized as in significant respects indigenes of still-traditional worlds. I don't mean here to challenge the conventional use of the term "indigenous," but only to point to how pre-modern ways of engaging with the world are alive and well inside modern societies.

6   For the latter, see, for example, Hessen, *Economic Roots of Newton's Principia*, on the fit of Newtonian projects with the economic needs of the emerging modern European society; Forman, "Behind Quantum Electronics," on how U.S. physics was shaped by national security needs in the World War II era; and Mirowski, "How Positivism Made a Pact," on the fit between three generations of the American philoso-

phy of science (Dewey, Reichenbach, and Kitcher) and dominant political/economic projects of the U.S. government in global politics.

7   For such primarily Western recognitions of this phenomenon, see Eisenstadt, *Multiple Modernities*; Giddens, "Living in a Post-Traditional Society."

8   Throughout this section and the next I draw on arguments from Felski, *Gender of Modernity*; Jardine, *Gynesis*; and C. V. Scott, GD. See also Bergeron, *Fragments of Development*, which was published too late for consideration here.

9   A confession: My father worked briefly in the 1920s for one of the early founders of time-budget studies (Frank Gilbreth). I still have amusing memories from several decades later of my mother's frustration upon returning home to discover that he had once again rearranged the kitchen appliances and furniture in order to reduce by a dozen seconds or so the time she spent getting from the refrigerator to the stove or the table to the sink. For the shift of midwifery skills to gynecologists and obstetricians, see Ehrenreich and English, *Witches, Midwives and Healers*; for the introduction of scientific rationality into housework, see Cowan, *More Work for Mother?*

10 Felski is particularly interesting on this point in her chapter 3, "Imagined Pleasures" in *Gender of Modernity*.

11 In addition to earlier feminist and postcolonial citations, see on this point such critics of the "underdevelopment" of modern Western epistemology as, for example, Beck and Nowotny et al., discussed in earlier chapters.

12 For influential feminist accounts of the strengths and limitations of Freudian theories, see Chodorow, *Reproduction of Mothering*; Dinnerstein, *Mermaid and the Minotaur*; Flax, *Thinking Fragments*.

13 Of course we should recollect that the original modernization and Marxian theorists were the great founders of sociology in the nineteenth century, before Freud's theories had appeared. In their theorizing they are oblivious to their own and other men's gender and sexual privileges, and certainly to issues of repressed sexual fears and desires. Moreover, the post–World War II era, when modernization theory was resurrected and reenergized, was one in which formal colonial rule was beginning to end and in which the second women's movement was about to gather steam in Europe and the United States. The economic, political, social, and psychic pre-conditions for the rise of postcolonial and feminist criticisms of Western and male-supremacist ideals of social progress were already in place as women and soon-to-be ex-colonials began to imagine futures for themselves which had been virtually inconceivable in preceding decades. To get a feel for

popular tastes in gender narratives in the post-war period, see Philip Wylie's *Generation of Vipers*. It was originally published in 1942 and was reprinted many times, including a "revised version" in 1978, which was quite a few years after his death.

14   The field of men's studies has provided illuminating studies here. See, e.g., Connell *Masculinities*; and Connell, Breines, and Eide, *Male Roles*.

9. MOVING ON

1   Recollect that exceptionalism holds that the West alone has developed the scientific and technological resources to achieve modernity and its social progress. Triumphalism holds that on balance modernity's history has had no truly dark side; that its association with disasters and atrocities such as Hiroshima or environmental destruction confuses the purely cognitive, technical core of research, for which science is responsible, with the political and social uses of the information such research produces, for which science bears no responsibility. As the Tom Lehrer song from the 1960s had rocket scientist Werner Von Braun singing, he was just responsible for getting the rockets up, not for where they came down.

2   Laura Nader (*Naked Science*) makes such a recommendation with respect to Western sciences and indigenous knowledge. Joan Kelly-Gadol borrows from W. E. B. Du Bois in pointing out that feminist historians, in assuming that women, too, are fully human, achieve a "double vision" of history with one eye on the standard accounts of progressive moments and the other on women's situation at those moments ("Social Relations of the Sexes"). A similar split vision appears in the familiar metaphors of the "outsider within," "border-lands," thinking "from margin to center," and standpoint theory's "starting research from marginalized lives."

3   There are additional research agendas which offer illuminating insights and promising directions for considering how to reconfigure scientific and technological research so that it advances progressive social transformation. Unfortunately, they are beyond the scope of this study. Here I am thinking especially of the feminist economics and international relations interrogations of the unity or coherence, inevitability, and power of "global capitalism" and of the narrow definitions in both "right" and "left" analyses of what should count as economic activity and international relations. Since scientific rationality and technical expertise are central to today's economic and international politics agendas, these accounts are especially fruitful. In addition to

the work of C. V. Scott (*Gender and Development*), considered in chapter 8, see, e.g., Bergeron, *Fragments of Development*; Gibson–Graham, *End of Capitalism*; Peterson, *Critical Rewriting*; and Sassen, *Globalization and Its Discontents.*

4 I thank David Hess for pointing out to me the importance of getting into focus more clearly the issue of where to go from here.

5 Recollect that these were the three sites of problems with standard histories, which came into view when one took women to be fully human, as Joan Kelly-Gadol put it in "Social Relations of the Sexes."

6 This is the second way Beck uses the term "reflexive." The first, which he does not always acknowledge using, is to call for modern institutions to undertake the same kinds of critical examination of their own conceptual frameworks and discourses as they typically bring to their objects of study. (This is a point I had made independently in *Is Science Multicultural?*)

7 For example, in Eisenstadt's *Multiple Modernities* and Eisenstadt's and Schluchter's *Early Modernities*. As indicated earlier, the multiple modernities arguments were also prefigured in Latour's arguments, along with many others from Northern science studies scholars.

8 See discussions of these agendas in the fields of men's studies and whiteness studies. See also Harding, "Reinventing Ourselves as Other," chap. 11 of *Whose Science?*; and "Can Men Be the Subjects of Feminist Thought?"

9 I say "middle-class" since poor women and wives of rich men have often worked outside the household in modern societies where the public-sphere vs. private-sphere conceptions are dominant.

10 Decades ago Heidi Hartmann's "The Family as the Locus of Gender, Class, and Political Struggle" demonstrated that the time women spend in household work increases by about nine hours per week if the household includes an adult (i.e., over age fourteen) male, and that this is so regardless of whether the woman works outside the household, there are children in the household, or the man contributes to taking on household responsibilities. I know of no recent data which counters such findings.

11 Nancy Folbre's recent study argues that contemporary data show that whoever has such household responsibilities will be handicapped in the public economy.

12 In the past, labor was represented as locationally fixed or stable and industries traveled to take advantage of it. Hence the "runaway" industries and the phenomenon of outsourcing manufacturing parts and services. This kind of labor relation certainly continues today. Yet it is also the case that labor now travels to where the work is; see Afshar

and Barrientos, *Women, Globalization, and Fragmentation*; Peterson, *Critical Rewriting of Global Political Economy*; Prugl, *Global Construction of Gender*; Sassen, *Globalization and Its Discontents*; Sparr, *Mortgaging Women's Lives*; Visvanathan et al., *Women, Gender and Development Reader*.

13   I follow the practice of feminists critically examining the lot of women in new forms of the global political economy to prefer the term "global restructuring" to "globalization." See note 12.

14   I do not mean to suggest that the situations and resources available to immigrant low-paid workers are the same or equally desirable as those available to professional women, but only that patterns of global restructuring are to be found in the organization of households and family relations no less than in economic, political, and public social relations.

15   What about men? Good question. These global political economy processes creating today's "new women" are also creating "new men." There have been some attempts to identify and understand diverse forms of transformations in masculinities at least, some of which are highly resistant to conventional male-supremacist ideals (Connell, Breines, and Eide, *Male Roles*). I cannot pursue this topic here beyond noting that the field of masculinity studies needs to be integrated as fully as possible into feminist studies.

16   In an older language one could ask why one should not start out from a central site of social and material reproduction to ask questions about projects of social and material production.

BIBLIOGRAPHY

— — — —

Ackerly, Brooke. *Political Theory and Feminist Social Criticism*. Cambridge: Cambridge University Press, 2000.

Adam, Barbara, Ulrich Beck, and Joost Van Loon. *The Risk Society and Beyond*. Thousand Oaks, Calif.: Sage, 2000.

Adas, Michael. *Machines as the Measure of Man*. Ithaca, N.Y.: Cornell University Press, 1989.

Afshar, Haleh, and Stephanie Barrientos, eds. *Women, Globalization, and Fragmentation in the Developing World*. New York: St. Martin's Press, 1999.

Agarwal, Bina. "The Gender and Environment Debate: Lessons from India." *Feminist Studies* 18, no. 1 (1993).

Aikenhead, Glen, ed. "Multicultural Sciences." Special issue, *Journal of Science Education* 85, no. 1 (2001).

Alexander, M. Jacqui, and Chandra Talpade Mohanty, eds. *Feminist Genealogies, Colonial Legacies, Democratic Futures*. New York: Routledge, 1997.

Almond, Gabriel A. and Sidney Verba. *The Civic Culture: Political Attitudes and Democracy in Five Nations*. Princeton: Princeton University Press, 1963.

American Association of University Women Educational Foundation. *Under the Microscope: A Decade of Gender Equity Projects in the Sciences*. Washington: AAUW Educational Foundation, 2004.

Amin, Samir. *Accumulation on a World Scale*. 2 vols. New York: Monthly Review Press, 1974.

——. *Delinking: Towards a Polycentric World*. New York: Zed Books, 1990.

——. *Eurocentrism.* New York: Monthly Review Press, 1989.

——. *Maldevelopment: Anatomy of a Global Failure.* New York: Zed Books, 1990.

Anzaldúa, Gloria. *Borderlands/La Frontera.* San Francisco: Spinsters/Aunt Lute, 1987.

Apffel-Marglin, Frederique, and Stephen Marglin. *Dominating Knowledge: Development, Culture, and Resistance.* Oxford: Clarendon Press, 1990.

Appleton, Helen, et al. "Claiming and Using Indigenous Knowledge." In *Missing Links,* ed. Gender Working Group. Ottawa: International Development Research Centre, Intermediate Technology Publications, and UNIFEM, 1995.

Ascher, Marcia. "Figures on the Threshold." In *Mathematics Elsewhere: An Exploration of Ideas across Cultures,* 161–90. Princeton: Princeton University Press, 2002.

Ashcroft, Bill, Gareth Griffiths, and Helen Tiffin, eds. *The Postcolonial Studies Reader.* New York: Routledge, 1995.

Audetat, Marc. "Re-Thinking Science, Re-Thinking Society." *Social Studies of Science* 31, no. 6 (2001): 950–56.

Balakrishnan, Gopal, ed. *Debating Empire.* New York: Verso, 2003.

Barad, Karen. "Agential Realism: Feminist Interventions in Understanding Scientific Practices." In *The Science Studies Reader,* ed. Mario Biagioli. New York: Routledge, 1999.

——. "Getting Real: Technoscientific Practices and the Materialization of Reality." *Differences* 10, no. 2 (1998): 87–128.

——. *Meeting the Universe Halfway: Quantum Physics and the Entanglement of Matter and Meaning.* Durham, N.C.: Duke University Press, 2007.

——. "Meeting the Universe Halfway: Realism and Social Constructivism Without Contradiction." In *Feminism, Science, and the Philosophy of Science,* ed. Lynn Hankinson Nelson and Jack Nelson. Dordrecht: Kluwer, 1996.

——. "Posthumanist Performativity: Toward an Understanding of How Matter Comes to Matter." *Signs: Journal of Women in Culture and Society* 28, no. 3 (2003): 801–32.

Barker, Drucilla, and Edith Kuiper. *Feminist Economics and the World Bank: History, Theory, and Politics.* New York: Routledge, 2006.

Basu, Amrita, Inderpal Grewal, Caren Kaplan, and Lisa Malkki, eds. "Globalization and Gender." Special issue, *Signs: Journal of Women in Culture and Society* 26, no. 4 (2001).

Beck, Ulrich. *The Reinvention of Politics: Rethinking Modernity in the Global Social Order.* Cambridge, U.K.: Polity Press, 1997.

——. "The Reinvention of Politics: Towards a Theory of Reflexive Modernization." In *Reflexive Modernization,* Ulrich Beck, Anthony Giddens, and Scott Lash. Cambridge, U.K.: Polity Press, 1994.

———. *Risk Society: Towards a New Modernity.* London: Sage, 1992. (German edition published in 1986.)

———. *World Risk Society.* Oxford: Blackwell, 1999.

Beck, Ulrich, and Elisabeth Beck-Gersheim. *The Normal Chaos of Love.* Cambridge, U.K.: Polity Press, 1995.

Beck, Ulrich, Anthony Giddens, and Scott Lash. *Reflexive Modernization: Politics, Tradition, and Aesthetics in the Modern Social Order.* Cambridge, U.K.: Polity Press, 1994.

Bellah, Robert N., ed. *Emile Durkheim on Morality and Society: Selections in Translation.* Chicago: University of Chicago Press, 1973.

Beneria, Lourdes. *Gender, Development, and Globalization: Economics as if All People Mattered.* New York: Routledge, 2003.

Benhabib, Seyla, ed. *Democracy and Difference: Contesting the Boundaries of the Political.* Princeton: Princeton University Press, 1996.

Bergeron, Suzanne. *Fragments of Development: Nation, Gender, and the Space of Modernity.* Ann Arbor: University of Michigan Press, 2006.

———. "Political Economy Discourses of Globalization and Feminist Politics." In "Globalization and Gender," special issue, *Signs: Journal of Women in Culture and Society* 26, no. 4 (2001): 983–1006.

Berman, Morris. *The Reenchantment of the World.* Ithaca, N.Y.: Cornell University Press, 1981.

Bhavnani, Kum-Kum, John Foran, and Priya Kurian, eds. *Feminist Futures: Re-imagining Women, Culture, and Development.* New York: Zed Books, 2003.

Biagioli, Mario. *The Science Studies Reader.* New York: Routledge, 1999.

Bielawski, Ellen. "Inuit Indigenous Knowledge and Science in the Arctic." In *Naked Science,* ed. Laura Nader. New York: Routledge, 1996.

Blaut, J. M. *The Colonizer's Model of the World: Geographical Diffusionism and Eurocentric History.* New York: Guilford Press, 1993.

Bloor, David. *Knowledge and Social Imagery.* London: Routledge and Kegan Paul, 1977.

Bordo, Susan. *The Flight to Objectivity.* Albany: State University of New York Press, 1987.

Boston Women's Health Collective. *Our Bodies, Ourselves.* Boston: New England Free Press, 1970. (Later editions published by Random House.)

Boyd, William. "Review of *Risk Society.*" *Economic Geography* 69, no. 4 (1993): 432–36.

Boyne, Roy. "The Politics of Risk Society." *History of the Human Sciences* 11, no. 3 (1998): 125–30.

Braidotti, Rosi, et al. *Women, the Environment, and Sustainable Development.* Atlantic Highlands, N.J.: Zed Books, 1994.

Brickhouse, Nancy. "Bringing in the Outsiders: Reshaping the Sciences of the Future." *Journal of Curriculum Studies* 26, no. 4 (1994): 401–16.

——. "Embodying Science: A Feminist Perspective on Learning." *Journal of Research in Science Teaching* 38, no. 3 (2001): 282–95.

Bridenthal, Renate, and Claudia Koonz. *Becoming Visible*. Boston: Houghton Mifflin, 1976.

Brockway, Lucille H. *Science and Colonial Expansion: The Role of the British Royal Botanical Gardens*. New York: Academic Press, 1979.

Bury, J. B. *The Idea of Progress: An Inquiry into Its Origin and Growth*. New York: Dover Publications, 1955 (1932).

Butler, Judith. *Antigone's Claim: Kinship Between Life and Death*. New York: Columbia University Press, 2000.

Cardoso, Fernando H., and Enzo Faletto. *Dependency and Development in Latin America*. Berkeley: University of California Press, 1979.

Castells, Manuel. *The Information Age: Economy, Society, and Culture*. Vols. 1–3. Oxford: Blackwell, 1996, 1997, 1998.

——. *The Power of Identity*. Vol. 2 of *The Information Age*. Oxford: Blackwell, 1997.

Caulfield, Mina Davis. "Imperialism, the Family, and Cultures of Resistance." *Socialist Review* 4, no. 2 (1974).

Cockburn, Cynthia. *Machinery of Dominance: Women, Men, and Technical Know-How*. London: Pluto Press, 1985.

Cockburn, Cynthia, and Susan Ormrod. *Gender and Technology in the Making*. London: Sage, 1993.

Code, Lorraine. *What Can She Know?* Ithaca, N.Y.: Cornell University Press, 1991.

Collins, Patricia Hill. *Black Feminist Thought: Knowledge, Consciousness, and the Politics of Empowerment*. New York: Routledge, 1991.

Connell, Robert. "Change among the Gatekeepers: Men, Masculinities, and Gender Equality in the Global Arena." *Signs: Journal of Women in Culture and Society* 30, no. 3 (2005): 1801–26.

——. *Masculinities*. Cambridge, U.K.: Polity Press, 1995.

Connell, Robert, Ingeborg Breines, and Ingrid Eide, eds. *Male Roles, Masculinities, and Violence: A Culture of Peace Perspective*. Paris: UNESCO, 2005.

Correll, Shelly J., Stephen Benard, and In Paik. "Getting a Job: Is There a Motherhood Penalty?" *American Journal of Sociology* 112 (2007): 1297–338.

Cowan, Ruth Schwartz. *More Work for Mother: The Ironies of Household Technology from the Open Hearth to the Microwave*. New York: Basic Books, 1983.

Crenshaw, Kimberle, et al. *Critical Race Theory*. Philadelphia: Temple University Press, 1995.

Crittenden, Ann. *The Price of Motherhood*. New York: Basic Books, 2001.

Crosby, Alfred. *The Columbian Exchange: Biological and Cultural Consequences of 1492.* Westport, Conn.: Greenwood Press, 1972.

——. *Ecological Imperialism: The Biological Expansion of Europe.* Cambridge: Cambridge University Press, 1987.

Cutliffe, Stephen H., and Carl Mitcham. *Visions of STS: Counterpoints in Science, Technology, and Society Studies.* Albany: State University of New York Press, 2001.

Danermark, Berth. "Review of *Re-Thinking Science.*" *Acta Sociologica* 46, no. 2 (2003): 166–76.

Davis, Angela. "The Black Woman's Role in the Community of Slaves." *Black Scholar* 2 (1971).

Dinnerstein, Dorothy. *The Mermaid and the Minotaur: Sexual Arrangements and Human Malaise.* New York: Harper and Row, 1976.

Dos Santos, Theotonio. "The Structure of Dependence." *American Economic Review* 60, no. 2 (1970): 231–36.

Draper, Elaine. "Risk, Society, and Social Theory." *Contemporary Sociology* 22, no. 5 (1993): 641–44.

Dupre, John. *The Disorder of Things: Metaphysical Foundations for the Disunity of Science.* Cambridge, Mass.: Harvard University Press, 1993.

——. "Metaphysical Disorder and Scientific Disunity." In *The Disunity of Science,* ed. Peter Galison and David J. Stump. Palo Alto, Calif.: Stanford University Press, 1996.

Ehrenreich, Barbara, and Deirdre English. *For Her Own Good.* New York: Doubleday, 1979.

Eisenstadt, S. N., ed. "Multiple Modernities." Special issue, *Daedalus* 129, no. 1 (2000). See esp. "Multiple Modernities," 1–30.

Eisenstadt, S. N., and Wolfgang Schluchter, eds. "Early Modernities." Special issue, *Daedalus* 127, no. 3 (1998).

Eisenstein, Zillah. *Capitalist Patriarchy and the Case for Socialist Feminism.* New York: Monthly Review Press, 1979.

Elam, Mark, and Oscar Juhlin. "When Harry Met Sandra." *Science As Culture* 7, no. 1 (1998): 95–109.

Elkana, Yehuda. "A Programmatic Attempt at an Anthropology of Knowledge." In *Sciences and Cultures, Sociology of the Sciences Yearbook,* ed. Everett Meldelsohn and Yehuda Elkana, Vol. 5, 1–76. Dordrecht: Reidel, 1991.

Escobar, Arturo. *Encountering Development: The Making and Unmaking of the Third World.* Princeton: Princeton University Press, 1995.

Etzkowitz, Henry, Carol Kemelgor, and Brian Uzzi. *Athena Unbound: The Advancement of Women in Science and Technology.* Cambridge: Cambridge University Press, 2000.

European Commission. *Waste of Talents: Turning Private Struggles into a*

Public Issue: Women and Science in the ENWISE *Countries.* Luxembourg: European Communities, 2003.

Fausto-Sterling, Anne. "The Bare Bones of Sex: Sex and Gender." *Signs: Journal of Women in Culture and Society* 30, no. 2 (2005): 1491–528.

——. *Myths of Gender: Biological Theories About Women and Men.* New York: Basic Books, 1994 (1985).

——. "Refashioning Race: DNA and the Politics of Health Care." *Differences: A Journal of Feminist Cultural Studies.* Forthcoming.

——. *Sexing the Body: Gender Politics and the Construction of Sexuality.* New York: Basic Books, 2000.

Feenberg, Andrew. *Alternative Modernity: The Technical Turn in Philosophy and Social Theory.* Berkeley: University of California Press, 1995.

——. *Transforming Technology: A Critical Theory Revisited.* New York: Oxford University Press, 2002.

Felski, Rita. *The Gender of Modernity.* Cambridge, Mass.: Harvard University Press, 1995.

Ferguson, Ann. *Blood at the Root: Motherhood, Sexuality, and Male Dominance.* New York: Pandora/Harper Collins, 1989.

Figueroa, Robert, and Sandra Harding, eds. *Science and Other Cultures: Issues in the Philosophy of Science and Technology.* New York: Routledge, 2003.

Flax, Jane. "Political Philosophy and the Patriarchal Unconscious: A Psychoanalytic Perspective on Epistemology and Metaphysics." In *Discovering Reality: Feminist Perspectives on Epistemology, Metaphysics, Methodology, and Philosophy of Science,* ed. Sandra Harding and Merrill B. Hintikka. Dordrecht: Kluwer, 2003 (1983).

——. "Postmodernism and Gender Relations in Feminist Theory." In *Feminism/Postmodernism,* ed. Linda Nicholson. New York: Routledge, 1990.

——. *Thinking Fragments: Psychoanalysis, Feminism, and Postmodernism in the Contemporary West.* Berkeley: University of California Press, 1990.

Folbre, Nancy. *The Invisible Heart: Economics and Family Values.* New York: New Press, 2001.

Forman, Paul. "Behind Quantum Electronics: National Security as Bases for Physical Research in the U.S., 1940–1960." *Historical Studies in Physical and Biological Sciences* 18 (1987).

Foucault, Michel. *Birth of the Clinic.* New York: Random House, 1994.

——. *Discipline and Punish.* Translated by Alan Sheridan. New York: Random House, 1977.

——. *Power/Knowledge: Selected Interviews and Other Writings 1972–77.* Edited by Colin Gordon. New York: Random House, 1980 (1972).

Frake, C. "The Ethnographic Study of Cognitive Systems." In *Anthropology and Human Behavior,* ed. T. Gladwin. Washington, D.C.: Anthropological Society of Washington, 1962.

Frank, Andre Gunder. *Capitalism and Underdevelopment in Latin America.* New York: Monthly Review Press, 1969.

Friedan, Betty. *The Feminine Mystique.* New York: W. W. Norton, 1963.

Friedman, Susan Stanford. "Definitional Excursions: The Meanings of Modern/Modernity/Modernism." *Modernism/Modernity* 8, no. 3 (2001): 493–513.

Fuller, Steve. "Review of *The New Production of Knowledge: The Dynamics of Science and Research in Contemporary Societies.*" *Sociology* 29, no. 1 (1995): 159–67.

——. *Science.* Buckingham, U.K.: Open University Press, 1997.

Furtado, Celso. *Economic Development of Latin America: A Survey from Colonial Times to the Cuban Revolution.* Cambridge: Cambridge University Press, 1970.

——. *The Myth of Economic Development.* Rio de Janeiro: Paz e Terra, 1974.

Galison, Peter. *How Experiments End.* Chicago: University of Chicago Press, 1987.

Galison, Peter, and David J. Stump, eds. *The Disunity of Science.* Palo Alto, Calif.: Stanford University Press, 1996.

Gender Working Group, United Nations Commission on Science and Technology for Development. *Missing Links: Gender Equity in Science and Technology for Development.* Ottawa: International Development Research Centre, Intermediate Technology Publications, and UNIFEM, 1995.

Gendzier, Irene. *Managing Political Change: Social Scientists and the Third World.* Boulder, Colo.: Westview Press, 1985.

Gibbons, Michael, et al. *The New Production of Knowledge: The Dynamics of Science and Research in Contemporary Societies.* Thousand Oaks, Calif.: Sage, 1994.

Gibson-Graham, J. K. *The End of Capitalism (As We Knew It): A Feminist Critique of Political Economy.* Oxford: Blackwell, 1996.

Giddens, Anthony. *The Consequences of Modernity.* Cambridge, U.K.: Polity Press, 1990.

——. "Living in a Post-Traditional Society." In *Reflexive Modernization,* Ulrich Beck, Anthony Giddens, and Scott Lash. Cambridge, U.K.: Polity Press, 1994.

——. *Modernity and Self-Identity.* Cambridge, U.K.: Polity Press, 1991.

——. *The Transformation of Intimacy.* Cambridge, U.K.: Polity Press, 1992.

Gill, Stephen. "Globalisation, Market Civilisation, and Disciplinary Neoliberalism." *Millennium: Journal of International Studies* 24, no. 3 (1995): 399–423.

Gilligan, Carol. *In a Different Voice.* Cambridge, Mass.: Harvard University Press, 1982.

Godin, Benoit. "Writing Performative History: The New *New Atlantis?*" *Social Studies of Science* 28, no. 3 (1998): 465–83.

Gole, Nilifer. "Global Expectations, Local Experiences: Non-Western Modernities." In *Through a Glass Darkly: Blurred Images of Cultural Tradition and Modernity over Distance and Time*, ed. Wil Arts. Leiden: Brill Publishers, 2000.

——. "Snapshots of Islamic Modernities." In "Multiple Modernities," ed. S. N. Eisenstadt, special issue, *Daedalus* 129, no. 1 (2000).

Goodenough, Ward H. "Navigation in the Western Carolines: A Traditional Science." In *Naked Science*, ed. Laura Nader. New York: Routledge, 1996.

Goonatilake, Susantha. *Aborted Discovery: Science and Creativity in the Third World*. London: Zed Books, 1984.

——. "A Project for Our Times." In *The Revenge of Athena*, ed. Z. Sardar. London: Mansell, 1988.

——. *Toward a Global Science: Mining Civilizational Knowledge*. Bloomington: Indiana University Press, 1998.

——. "The Voyages of Discovery and the Loss and Rediscovery of the 'Other's' Knowledge." *Impact of Science on Society*, no. 167 (1992): 241–64.

Gross, Paul R., and Norman Levitt. *Higher Superstition: The Academic Left and Its Quarrels with Science*. Baltimore: Johns Hopkins University Press, 1994.

Gross, Paul R., and Martin W. Lewis, eds. *The Flight from Science and Reason. Annals of the New York Academy of Sciences*, no. 775 (1996).

Gunew, Sneja. *Haunted Nations: The Colonial Dimensions of Multiculturalism*. New York: Routledge, 2004.

Hacking, Ian. "The Disunities of the Sciences." In *The Disunity of Science*, ed. Peter Galison and David J. Stump. Palo Alto, Calif.: Stanford University Press, 1996.

——. *Representing and Intervening*. Cambridge: Cambridge University Press, 1983.

Hagstrom, Warren O. *The Scientific Community*. New York: Basic Books, 1965.

Hall, J. R. "Review of *Risk Society*." *The Sociological Review* 42 (1994): 344–46.

Hammonds, Evelynn. *The Logic of Difference: A History of Race in Science and Medicine in the United States*. Forthcoming.

Haraway, Donna. *Modest—Witness@ Second—Millennium.FemaleMan©—Meets —OncoMouse™: Feminism and Technoscience*. New York: Routledge, 1997.

——. *Primate Visions: Gender, Race, and Nature in the World of Modern Science*. New York: Routledge, 1989.

———. *Simians, Cyborgs, and Women: The Reinvention of Nature.* New York: Routledge, 1991.

———. "Situated Knowledges: The Science Question in Feminism and the Privilege of Partial Perspectives." In *Simians, Cyborgs, and Women.* New York: Routledge, 1991.

Harcourt, Wendy, ed. *Feminist Perspectives on Sustainable Development.* London: Zed Books, 1994.

Harding, Sandra. "After the Neutrality Ideal: Science, Politics, and 'Strong Objectivity.'" *Social Research* 59 (1992): 567–87.

———. "Difference and Power in Feminist Epistemology and Science Studies." *Hypatia: A Journal of Feminist Philosophy.* Forthcoming.

———. "Gender, Development, and Post-Enlightenment Philosophies of Science." In *Decentering the Center,* ed. S. Harding and Uma Narayan. Bloomington: Indiana University Press, 2000.

———. "Gender, Modernity, Knowledge: Postcolonial Standpoints." Chap. 7 in *Is Science Multicultural? Postcolonialisms, Feminisms, and Epistemologies.* Bloomington: Indiana University Press, 1998.

———. "The Instability of the Analytical Categories of Feminist Theory." *Signs: Journal of Women in Culture and Society* 11, no. 4 (1986): 645–64.

———. "Is Modern Science an Ethnoscience?" In *Sociology of the Sciences Yearbook,* ed. T. Shinn, J. Spaapen, and Raoul Waast. Dordrecht: Kluwer, 1996.

———. "Is Science Multicultural? Challenges, Resources, Opportunities, Uncertainties." In *Configurations* 2, no. 2 (1994); and *Multiculturalism: A Reader,* ed. David Theo Goldberg. London: Blackwell, 1994.

———. *Is Science Multicultural? Postcolonialisms, Feminisms, and Epistemologies.* Bloomington: Indiana University Press, 1998.

———. "Modernity, Science, and Democracy." In *Modernity in Transit/La Modernité en Transit,* ed. Pascal Gin and Walter Moser. Toronto: Les Editions du GREF, 2006; and *Social Philosophy Today,* vol. 22: *Science, Technology, and Social Justice,* ed. John Rowan. Charlottesville, Va.: Philosophy Documentation Center, 2007.

———. "Multicultural and Global Feminist Philosophies of Science: Resources and Challenges." In *Feminism, Science, and the Philosophy of Science,* ed. Lynn Hankinson Nelson and Jack Nelson. Dordrecht: Kluwer, 1996.

———. "Rethinking Standpoint Epistemology." *Feminist Epistemologies,* ed. L. Alcoff and E. Potter. New York: Routledge, 1992.

———. *Science and Social Inequality: Feminist and Postcolonial Issues.* Champaign: University of Illinois Press, 2006.

———. *The Science Question in Feminism.* Ithaca, N.Y.: Cornell University Press, 1986.

——. "Two Influential Theories of Ignorance and Philosophy's Interests in Ignoring Them." *Hypatia: A Journal of Feminist Philosophy* 21, no. 3 (2006): 20–36.

——. "Why Has the Sex/Gender System Become Visible Only Now?" In *Discovering Reality*, ed. Sandra Harding and M. Hintikka. Dordrecht: Reidel, 2003 (1983).

——. *Whose Science? Whose Knowledge? Thinking from Women's Lives.* Ithaca, N.Y.: Cornell University Press, 1991.

——, ed. *Can Theories Be Refuted? Essays on the Duhem-Quine Thesis.* Dordrecht: Reidel, 1976.

——, ed. *Feminism and Methodology: Social Science Issues.* Bloomington: Indiana University Press, 1987.

——, ed. *The Feminist Standpoint Theory Reader: Intellectual and Political Controversies.* New York: Routledge, 2004.

——, ed. *The "Racial" Economy of Science: Toward a Democratic Future.* Bloomington: Indiana University Press, 1993.

Harding, Sandra, and Merrill Hintikka, eds. *Discovering Reality: Feminist Perspectives on Epistemology, Metaphysics, Methodology and Philosophy of Science.* 2nd ed. Dordrecht: Kluwer, 2003 (1983).

Harding, Sandra, and Elizabeth McGregor. "The Gender Dimension of Science and Technology." In *UNESCO World Science Report*, ed. Howard J. Moore. Paris: UNESCO, 1996.

Harding, Sandra, and Uma Narayan, eds. *Decentering the Center: Philosophy for a Multicultural, Postcolonial, and Feminist World.* Bloomington: Indiana University Press, 2000.

Harding, Sandra, and Kate Norberg, eds. "New Feminist Approaches to Social Science Methodologies." Special issue, *Signs: Journal of Women in Culture and Society* 30, no. 4 (2005).

Harding, Sandra, and Jean O'Barr, eds. *Sex and Scientific Inquiry.* Chicago: University of Chicago Press, 1987.

Hardt, Michael, and Antonio Negri. *Empire.* Cambridge, Mass.: Harvard University Press, 2000.

——. *Multitude: War and Democracy in the Age of Empire.* New York: Penguin Press, 2004.

Harris, Steve. "Long-Distance Corporations and the Geography of Natural Knowledge." *Configurations* 6, no. 2 (1998).

Hartmann, Heidi. "The Family as the Locus of Gender, Class, and Political Struggle: The Example of Housework." *Signs: Journal of Women in Culture and Society* 6, no. 3 (1981).

——. "The Unhappy Marriage of Marxism and Feminism." In *Women and Revolution*, ed. Lydia Sargent. Boston: South End Press, 1981.

Hartsock, Nancy. "The Feminist Standpoint: Developing the Ground for a Specifically Feminist Historical Materialism." In *Discovering Reality: Feminist Perspectives on Epistemology, Metaphysics, Methodology, and Philosophy of Science*, ed. Sandra Harding and Merrill Hintikka. Dordrecht: Reidel/Kluwer, 1983.

Harvey, David. *The Condition of Postmodernity.* Oxford: Blackwell, 1989.

Hayles, N. Katherine. "Constrained Constructivism: Locating Scientific Inquiry in the Theater of Representation." *Realism and Representation*, ed. George Levine. Madison: University of Wisconsin Press, 1993.

Headrick, Daniel R., ed. *The Tools of Empire: Technology and European Imperialism in the Nineteenth Century.* New York: Oxford University Press, 1981.

Hecht, Gabrielle, and Warwick Anderson, eds. "Postcolonial Technoscience." Special issue, *Social Studies of Science* 32, no. 5–6 (2002).

Heilbron, Johan, Lars Magnusson, and Bjorn Wittrock. *The Rise of the Social Sciences and the Formation of Modernity.* Dordrecht: Kluwer, 1998.

Held, David, and Anthony McGrew, eds. *The Global Transformations Reader.* Cambridge, U.K.: Polity Press, 2000.

Hess, David J. *Science and Technology in a Multicultural World: The Cultural Politics of Facts and Artifacts.* New York: Columbia University Press, 1995.

Hesse, Mary. *Models and Analogies in Science.* Notre Dame, Ind.: Notre Dame Press, 1966.

Hessen, Boris. *The Economic Roots of Newton's Principia.* New York: Howard Fertig, 1971.

Hobson, John M. *The Eastern Origins of Western Civilisation.* Cambridge: Cambridge University Press, 2004.

Hollinger, David. *Science, Jews, and Secular Culture.* Princeton: Princeton University Press, 1996.

hooks, bell. "Choosing the Margin as a Space of Radical Openness." In *Yearning: Race, Gender, and Cultural Politics*, 145–53. Boston: South End Press, 1990.

Horton, Robin. "African Traditional Thought and Western Science." Parts 1 and 2. *Africa* 37 (1967).

Hubbard, Ruth. *The Politics of Women's Biology.* New Brunswick, N.J.: Rutgers University Press, 1990.

Hubbard, Ruth, M. S. Henifin, and Barbara Fried, eds. *Biological Woman: The Convenient Myth.* Cambridge, Mass.: Schenkman, 1982.

Hutchins, Edwin. *Cognition in the Wild.* Cambridge, Mass.: MIT Press, 1996.

Ihde, Don. *Philosophy of Technology: An Introduction.* St. Paul, Minn.: Paragon House, 1993.

——. *Technology and the Lifeworld.* Bloomington: Indiana University Press, 1990.

Inayatullah, Naeem, and David L. Blaney. *International Relations and the Problem of Difference*. New York: Routledge, 2004.

Indigenous Knowledge and Development Monitor. Online. http://www.nuffic .ni/ciran/ikdm.html.

Inkeles, Alex, and David H. Smith. *Becoming Modern: Individual Change in Six Developing Countries*. Cambridge, Mass.: Harvard University Press, 1974.

Jacob, Margaret. *The Cultural Meanings of the Scientific Revolution*. New York: Knopf, 1988.

Jaggar, Alison. "Feminist Politics and Epistemology." In *Feminist Politics and Human Nature*. Totowa, N.J.: Rowman and Littlefield, 1988.

———. "Love and Knowledge: Emotion in Feminist Epistemology." In *Gender/Body/Knowledge*, ed. Susan Bordo and Alison Jaggar. New Brunswick, N.J.: Rutgers University Press, 1989.

James, C. L. R. *The Black Jacobins*, 2nd ed. rev. New York: Vintage, 1963.

Jameson, Fredric. " 'History and Class Consciousness' as an Unfinished Project." *Rethinking Marxism* 1 (1988): 49–72.

———. *The Political Unconscious: Narrative as a Socially Symbolic Act*. Ithaca, N.Y.: Cornell University Press, 1981.

———. *Postmodernism, or, The Cultural Logic of Late Capitalism*. Durham, N.C.: Duke University Press, 1991.

Jardine, Alice. *Gynesis: Configurations of Woman and Modernity*. Ithaca, N.Y.: Cornell University Press, 1985.

Jasanoff, Sheila. *Designs on Nature: Science and Democracy in Europe and the United States*. Princeton: Princeton University Press, 2005.

———., ed. *States of Knowledge: The Co-Production of Science and Social Order*. New York: Routledge, 2004.

Joseph, George Gheverghese. *The Crest of the Peacock: Non-European Roots of Mathematics*. New York: I. B. Tauris, 1991.

Kamenka, Eugene, ed. *The Portable Karl Marx*. New York: Viking Press, 1983.

Kazanjian, Arminee. *Doing the Right Thing, Not Just Doing Things Right: A Framework for Decisions about Technology*. Ottawa: International Development Research Centre, 1995.

Keller, Evelyn Fox. *A Feeling for the Organism*. San Francisco: Freeman, 1983.

———. *Reflections on Gender and Science*. New Haven, Conn.: Yale University Press, 1984.

———. *Secrets of Life, Secrets of Death: Essays on Language, Gender, and Science*. New York: Routledge, 1992.

Keller, Evelyn Fox, and Helen E. Longino, eds. *Feminism and Science*. New York: Oxford University Press, 1996.

Kelly, Alison, ed. *The Missing Half: Girls and Science Education.* Manchester, U.K.: Manchester University Press, 1981.

——, ed. *Science for Girls?* Philadelphia: Open University Press, 1987.

Kelly-Gadol, Joan. "Did Women Have a Renaissance?" In *Becoming Visible,* ed. R. Bridenthal and C. Koonz. Boston: Houghton Mifflin, 1976.

——. "The Social Relations of the Sexes: Methodological Implications of Women's History." *Signs: Journal of Women in Culture and Society* 1, no. 4 (1976): 810–23. 1987.

Kettel, Bonnie. "Key Paths for Science and Technology." In *Missing Links,* ed. Gender Working Group. Ottawa: International Development Research Centre, Intermediate Technology Publications, and UNIFEM, 1995.

Khor, Kok Peng. "Science and Development: Underdeveloping the Third World." In *The Revenge of Athena,* ed. Z. Sardar. London: Mansell, 1988.

Kimball, Roger. *Tenured Radicals.* New York: Ivan Dee, 1998.

Klein, Hans K., and Daniel Lee Kleinman. "The Social Construction of Technology: Structural Considerations." *Science, Technology and Human Values* 27, no. 1 (2002): 28–52.

Kleinman, Daniel Lee, and Steven P. Vallas, "Science, Capitalism, and the Rise of the 'Knowledge Worker': The Changing Structure of Knowledge Production in the United States." *Theory and Society* 30 (2001): 451–92.

Kline, Morris. *Mathematics: The Loss of Certainty.* New York: Oxford, 1980.

Knorr-Cetina, Karin. *The Manufacture of Knowledge: An Essay on the Constructivist and Contextual Nature of Knowledge.* New York: Pergamon, 1981.

Koblitz, Ann Hibner. "Challenges in Interpreting Data." In "The Gender Dimension of Science and Technology," ed. Sandra Harding and Elizabeth McGregor, *World Science Report 1996.* Paris: UNESCO, 1995.

Kochhar, R. K. "Science in British India." Parts 1 and 2. *Current Science* (India) 63, no. 11 (1992): 689–94; 64, no. 1 (1993): 55–62.

Kuhn, Thomas S. *The Structure of Scientific Revolutions.* 2nd ed. Chicago: University of Chicago Press, 1970 (1962).

Kumar, Deepak. "Problems in Science Administration: A Study of the Scientific Surveys in British India 1757–1900." In *Science and Empires: Historical Studies About Scientific Development and European Expansion,* ed. Patrick Petitjean et al. Dordrecht: Kluwer, 1992.

——. *Science and Empire: Essays in Indian Context (1700–1947).* Delhi: Anamika Prakashan, and National Institute of Science, Technology and Development, 1991.

Lach, Donald F. *Asia in the Making of Europe.* Vol. 2. Chicago: University of Chicago Press, 1977.

Lakatos, Imre, and Alan Musgrave., eds. *Criticism and the Growth of Knowledge.* New York: Cambridge University Press, 1970.

Latour, Bruno. *The Pasteurization of France.* Cambridge, Mass.: Harvard University Press, 1988.

——. *Politics of Nature: How to Bring the Sciences into Democracy.* Tr. Catherine Porter. Cambridge, Mass: Harvard University Press, 2004.

——. *Science in Action.* Cambridge, Mass.: Harvard University Press, 1987.

——. *We Have Never Been Modern.* Trans. Catherine Porter. Cambridge, Mass.: Harvard University Press, 1993.

Latour, Bruno, and Steve Woolgar. *Laboratory Life: The Social Construction of Scientific Facts.* Beverly Hills, Calif.: Sage, 1979.

Lave, Jean. "The Savagery of the Domestic Mind." In *Naked Science,* ed. Laura Nader. New York: Routledge, 1996.

Lerner, Daniel. *The Passing of Traditional Society: Modernizing the Middle East.* New York: Free Press, 1964 (1958).

Linder, Fletcher, and Joe Spear. "Review of *Rethinking Knowledge.*" *Contemporary Sociology* 32, no. 1 (2003): 255–57.

Linklater, Andrew. "Review of *The Reinvention of Politics.*" *Sociological Review* 45, no. 4 (1997): 731–34.

Lionnet, Françoise, et al. "Development Cultures." Special issue, *Signs: Journal of Women in Culture and Society* 29, no. 2 (2004).

Lionnet, Françoise, and Shiu-mei Shih, eds. *Minority Transnationalisms.* Durham, N.C.: Duke University Press, 2005.

Livingstone, David. N. *Putting Science in Its Place: Geographies of Scientific Knowledge.* Chicago: University of Chicago Press, 2003.

Lloyd, Genevieve. *The Man of Reason: "Male" and "Female" in Western Philosophy.* Minneapolis: University of Minnesota Press, 1984.

Longino, Helen. *The Fate of Knowledge.* Princeton: Princeton University Press, 2002.

——. *Science as Social Knowledge.* Princeton: Princeton University Press, 1990.

Machlup, Fritz. *The Production and Distribution of Knowledge in the United States.* Princeton: Princeton University Press, 1976.

MacKenzie, Donald, and Judy Wajcman. *The Social Shaping of Technology,* 2nd ed. Buckingham, U.K.: Open University Press, 1999.

MacKinnon, Catharine A. "Feminism, Marxism, Method, and the State: Toward Feminist Jurisprudence." *Signs: Journal of Women in Culture and Society* 8, no. 4 (1983): 635–58.

Maffie, James. " 'To Think with a Good Heart.' " *Hypatia: A Journal of Feminist Philosophy.* Forthcoming.

——. "To Walk in Balance: An Encounter Between Contemporary Western Science and Conquest-Era Nahua Philosophy." In *Science and Other Cultures: Issues in Philosophies of Science and Technology,* ed. Robert Figueroa and Sandra Harding. New York: Routledge, 2003.

——, ed. "Science, Modernity, Critique." Special issue on Meera Nanda's *Prophets Facing Backward: Postmodern Critiques of Science and Hindu Nationalism in India, Social Epistemology* 19, no. 1 (2005).

——, ed. "Truth from the Perspective of Comparative World Philosophy." Special issue, *Social Epistemology* 15, no. 4 (2001).

Malinowski, Bronislaw. *Magic, Science, and Religion and Other Essays.* Garden City, N.Y.: Doubleday Anchor, 1948 (1925).

Marchand, Marianne H., and Anne Sisson Runyan, eds. *Gender and Global Restructuring.* New York: Routledge, 2000.

Marchessault, Janine, and Kim Sawchuk. *Wild Science: Reading Feminism, Medicine, and the Media.* London: Routledge, 2000.

Marshall, Barbara. L. "Critical Theory, Feminist Theory, and Technology Studies." In *Modernity and Technology,* ed. Thomas J. Misa, Philip Brey, and Andrew Feenberg. Cambridge, Mass.: MIT Press, 2003.

Massachusetts Institute of Technology. "A Study on the Status of Women Faculty at MIT." *MIT Faculty Newsletter* 11, no. 4 (March 1999). Online.

Maxwell, Linda, Karen Slavin, and Kerry Young, eds. *Gender and Research.* Brussels: European Commission, 2002.

Mayberry, Maralee, Banu Subramaniam, and Lisa H. Weasel, eds. *Feminist Science Studies: A New Generation.* New York: Routledge, 2001.

McClellan, James E. *Colonialism and Science: Saint Domingue in the Old Regime.* Baltimore: Johns Hopkins University Press, 1992.

McClintock, Anne. *Imperial Leather: Race, Gender, and Sexuality in the Colonial Contest.* New York: Routledge, 1995.

McCumber, John. *Time in the Ditch: American Philosophy and the McCarthy Era.* Evanston, Ill.: Northwestern University Press, 2000.

Merchant, Carolyn. *The Death of Nature: Women, Ecology, and the Scientific Revolution.* New York: Harper and Row, 1980.

Mertes, Tom, ed. *A Movement of Movements: Is Another World Really Possible?* New York: Verso, 2004.

Mies, Maria. *Patriarchy and Accumulation on a World Scale: Women in the International Division of Labor.* Atlantic Highlands, N.J.: Zed Books, 1986.

Millman, Marcia, and Rosabeth Moss Kanter, eds. Introduction. *Another Voice: Feminist Perspectives on Social Life and Social Science.* New York: Doubleday Anchor Books, 1975.

Mirowski, Philip. "How Positivism Made a Pact with the Postwar Social Sciences in the United States." In *The Politics of Method in the Human Sciences: Positivism and Its Epistemological Others,* ed. George Steinmetz. Durham, N.C.: Duke University Press, 2005.

Misa, Thomas J., Philip Brey, and Andrew Feenberg, eds. *Modernity and Technology.* Cambridge, Mass.: MIT Press, 2003.

Mohanty, Chandra Talpade. *Feminism Without Borders: Decolonizing Theory, Practicing Solidarity.* Durham, N.C.: Duke University Press, 2003.

——. "Under Western Eyes: Feminist Scholarship and Colonial Discourses." In *Third World Women and the Politics of Feminism*, ed. Chandra Talpade Mohanty, Ann Russo, and Lourdes Torres. Bloomington: Indiana University Press, 1991.

Moraze, Charles, ed. *Science and the Factors of Inequality.* Paris: UNESCO, 1979.

Morrison, Toni. *Beloved.* New York: Plume Publishing, 1998.

Muckerjee, Ramkrishna. *The Rise and Fall of the British East India Company.* New York: Monthly Review Press, 1974.

Mullin, Katherine. "Modernisms and Feminisms." In *The Cambridge Companion to Feminist Literary Theory*, ed. Ellen Rooney. New York: Cambridge University Press, 2006.

Muntemba, Shimwaayi, and Rivimbo Chimedza. "Women Spearhead Food Security." In *Missing Links*, ed. Gender Working Group. Ottawa: International Development Research Centre, Intermediate Technology Publications, and UNIFEM, 1995.

Murata, Junichi. "Creativity of Technology and the Modernization Process of Japan." In *Science and Other Cultures: Issues in Philosophies of Science and Technology*, ed. Robert Figueroa and Sandra Harding. New York: Routledge, 2003.

Nader, Laura, ed. *Naked Science: Anthropological Inquiry into Boundaries, Power, and Knowledge.* New York: Routledge, 1996.

Nanda, Meera. *Prophets Facing Backward: Postmodern Critiques of Science and Hindu Nationalism in India.* New Brunswick, N.J.: Rutgers University Press, 2004.

Nandy, Ashis. *The Intimate Enemy: Loss and Recovery of Self under Colonialism.* Delhi: Oxford University Press, 1983.

——, ed. *Science, Hegemony, and Violence: A Requiem for Modernity.* Delhi: Oxford University Press, 1990.

Narayan, Uma. *Dislocating Cultures: Identities, Traditions, and Third World Feminism.* New York: Routledge, 1997.

——. "The Project of a Feminist Epistemology: Perspectives from a Nonwestern Feminist." In *Gender/Body/Knowledge*, ed. Susan Bordo and Alison Jaggar. New Brunswick, N.J.: Rutgers University Press, 1989.

National Science Foundation. *New Formulas for America's Workforce: Girls in Science and Engineering.* Washington: National Science Foundation Education and Human Resources Directorate, 2004.

Needham, Joseph. "The Laws of Man and the Laws of Nature." In *The Grand Titration: Science and Society in East and West.* Toronto: University of Toronto Press, 1969.

——. *Science and Civilisation in China*. 7 vols. Cambridge: Cambridge University Press, 1956–2004.

New, Caroline. "Class Society or Risk Society?" *Radical Philosophy* 66 (1994): 56.

Noble, David. *The Religion of Technology*. New York: Knopf, 1995.

Nowotny, Helga, Peter Scott, and Michael Gibbons. *Re-Thinking Science: Knowledge and the Public in an Age of Uncertainty*. Cambridge, U.K.: Polity Press, 2001.

O'Brien, Mary. *The Politics of Reproduction*. New York: Routledge & Kegan Paul, 1981.

Oldenziel, Ruth. *Making Technology Masculine: Women, Men, and the Machine in America, 1880–1945*. Amsterdam: Amsterdam University Press, 1999.

Oldenziel, Ruth, Annie Canel, and Karin Zachmann, eds. *Building Bridges, Crossing Boundaries: Comparing the History of Women Engineers*. London: Harwood, 2000.

O'Neill, Eileen. "Invisible Ink: Early Modern Women Philosophers and Their Fate in History." In *Feminist Philosophies*, ed. Janet Kourany. New York: Prentice Hall, 1998.

Ortiz, Renato. "From Incomplete Modernity to World Modernity." In "Multiple Modernities," ed. S. N. Eisenstadt, special issue, *Daedalus* 129, no. 1 (2000): 249–60.

Oyewumi, Oyeronke. *The Invention of Women: Making an African Sense of Western Gender Discourses*. Minneapolis: University of Minnesota Press, 1997.

Parsons, Talcott. *The Evolution of Societies*. Englewood Cliffs, N.J.: Prentice Hall, 1977.

Passavant, Paul A., and Jodi Dean, eds. *Empire's New Clothes: Reading Hardt and Negri*. New York: Routledge, 2004.

Pateman, Carole. *The Sexual Contract*. Palo Alto, Calif.: Stanford University Press, 1988.

Pels, Dick. "Strange Standpoints, or How to Define the Situation for Situated Knowledge." In *The Intellectual as Stranger*. New York: Routledge, 2001.

Peterson, V. Spike. *A Critical Rewriting of Global Political Economy: Integrating Reproductive, Productive, and Virtual Economies*. London: Routledge, 2003.

Peterson, V. Spike, and Anne Sisson Runyan. *Global Gender Issues*. Boulder, Colo.: Westview Press, 1993.

Petitjean, Patrick, et al., eds. *Science and Empires: Historical Studies About Scientific Development and European Expansion*. Dordrecht: Kluwer, 1992.

Philip, Kavita. *Civilising Natures: Race, Resources, and Modernity in Colonial South India*. New Delhi: Orient Longman, 2003.

Pieterse, Jan Nederveen. *Globalization or Empire?* New York: Routledge, 2004.

Plumwood, Val. *Feminism and the Mastery of Nature.* New York: Routledge, 1993.

Popper, Karl. *Conjectures and Refutations: The Growth of Scientific Knowledge.* 4th ed. London: Routledge & Kegan Paul, 1972.

Potter, Elizabeth. *Feminism and Philosophy of Science: An Introduction.* New York: Routledge, 2006.

——. *Gender and Boyle's Law of Gases.* Bloomington: Indiana University Press, 2001.

Prakash, Gyan. *Another Reason: Science and the Imagination of Modern India.* Princeton: Princeton University Press, 1999.

Prasad, Amit. "Beyond Modern vs. Alternative Science Debate: Analysis of Magnetic Resonance Imaging Research." *Economic and Political Weekly* (India), 21 January 2006.

Pratt, Mary Louise. *Imperial Eyes: Travel Writing and Transculturation.* New York: Routledge, 1992.

Price, Derek. *Little Science, Big Science.* New York: Columbia University Press, 1963.

Proctor, Robert. *Racial Hygiene: Medicine Under the Nazis.* Cambridge, Mass.: Harvard University Press, 1988.

——. *Value-Free Science? Purity and Power in Modern Knowledge.* Cambridge, Mass.: Harvard University Press, 1991.

Prugl, Elisabeth. *The Global Construction of Gender: Home-Based Work in the Political Economy of the 20th Century.* New York: Columbia University Press, 1999.

Pyenson, Lewis. *Cultural Imperialism and Exact Sciences.* New York: Peter Lang, 1985.

Quine, W. V. O. "Two Dogmas of Empiricism." In *From a Logical Point of View.* Cambridge, Mass: Harvard University Press, 1953.

——. *Word and Object.* Cambridge, Mass.: MIT Press, 1960.

Quine, W. V. O., and J. S. Ullian. *The Web of Belief.* New York: McGraw Hill, 1978.

Raffensperger, Carolyn, and Joel Tickner, eds. *Protecting Public Health and the Environment: Implementing the Precautionary Principle.* Washington, D.C.: Island Press, 1999.

Ravetz, Jerome. *Scientific Knowledge and Its Social Problems.* Oxford: Clarendon Press, 1971.

Reekie, Gail. *Temptations: Sex, Selling, and the Department Store.* Sydney: Allen and Unwin, 1993.

Reingold, Nathan, and Marc Rothenberg, eds. *Scientific Colonialism: Cross-Cultural Comparisons.* Washington: Smithsonian Institution Press, 1987.

Reisch, George A. *How the Cold War Transformed Philosophy of Science*. New York: Cambridge University Press, 2005.

Restivo, Sal. *Mathematics in Society and History: Sociological Inquiries*. Dordrecht: Kluwer, 1992.

——. "Politics of Latour." *Organization and Environment* 18, no. 1 (2005): 111–15.

Riesebrodt, M. *Pious Passion: The Emergence of Modern Fundamentalism in the United States and Iran*. Berkeley: University of California Press, 1993.

Rip, Arie. "Reflections on the Transformation of Science." *Metascience* 11, no. 3 (2002): 317–23.

Robertson, Roland. "Review of *The Reinvention of Politics*." *American Journal of Sociology* 103, no. 6 (1998): 1734–35.

Robinson, William I. *Promoting Polyarchy: Globalization, U.S. Intervention, and Hegemony*. New York: Cambridge University Press, 1996.

Rodney, Walter. *How Europe Underdeveloped Africa*. Washington: Howard University Press, 1982.

Rorty, Richard. *Philosophy and the Mirror of Nature*. Princeton: Princeton University Press, 1979.

Rose, Hilary. "Hand, Brain, and Heart: A Feminist Epistemology for the Natural Sciences." *Signs: Journal of Women in Culture and Society* 9, no. 1 (1983).

Rose, Hilary, and Helga Nowotny. "Countermovements in the Sciences: The Sociology of Alternatives to Big Science." *Sociology of the Sciences Yearbook*. Dordrecht: Kluwer, 1979.

Rose, Hilary, and Steven Rose. "The Incorporation of Science." In *Ideology of/in the Natural Sciences*, ed. H. and S. Rose. Cambridge, Mass.: Schenkman, 1976.

Ross, Andrew. *The Science Wars*. Durham, N.C.: Duke University Press, 1996.

Rossiter, Margaret. *Women Scientists in America: Before Affirmative Action*. Baltimore: Johns Hopkins University Press, 1995.

——. *Women Scientists in America: Struggles and Strategies to 1940*. Baltimore: Johns Hopkins University Press, 1982.

Rostow, W. W. *The Stages of Economic Growth*. Cambridge: Cambridge University Press, 1960.

Rouse, Joseph. "Barad's Feminist Naturalism." *Hypatia: A Journal of Feminist Philosophy* 19, no. 1 (2004): 142–61.

——. *Engaging Science: How to Understand Its Practices Philosophically*. Ithaca, N.Y.: Cornell University Press, 1996.

——. "Feminism and the Social Construction of Scientific Knowledge." In *Feminism, Science, and the Philosophy of Science*, ed. Lynn Hankinson Nelson and Jack Nelson. Dordrecht: Kluwer, 1996.

———. *How Scientific Practices Matter: Reclaiming Philosophical Naturalism.* Chicago: University of Chicago Press, 2002.

———. *Knowledge and Power: Toward a Political Philosophy of Science.* Ithaca, N.Y.: Cornell University Press, 1987.

Ruddick, Sara. *Maternal Thinking.* Boston: Beacon Press, 1989.

Runciman, W. G., ed. *Max Weber: Selections in Translation.* Cambridge: Cambridge University Press, 1978.

Sabra, I. A. "The Scientific Enterprise." In *The World of Islam,* ed. B. Lewis. London: Thames and Hudson, 1976.

Sachs, Wolfgang, ed. *The Development Dictionary: A Guide to Knowledge as Power.* Atlantic Highlands, N.J.: Zed Books, 1992.

Said, Edward. *Orientalism.* New York: Pantheon, 1978.

Salomon, Jean-Jacques. "Society Talks Back." *Nature* 412, no. 9 (2001): 585–86.

Sandoval, Chela. "U.S. Third World Feminism: The Theory and Method of Oppositional Consciousness in the Postmodern World." *Genders,* no. 10 (1991): 1–24.

Sardar, Ziauddin, ed. "Islamic Science: The Contemporary Debate." In *Encyclopedia of the History of Science, Technology, and Medicine in Non-Western Cultures,* ed. Helaine Selin, 455–58. Dordrecht: Kluwer, 1997.

———. *The Revenge of Athena: Science, Exploitation, and the Third World.* London: Mansell, 1988.

Sargent, Lydia, ed. *Women and Revolution: A Discussion of the Unhappy Marriage of Marxism and Feminism.* Boston: South End Press, 1981.

Sassen, Saskia. *Globalization and Its Discontents.* Chaps. 7–8. New York: New Press, 1998.

Sayre, Anne. *Rosalind Franklin and DNA.* New York: W. W. Norton, 1975.

Schiebinger, Londa. *Has Feminism Changed Science?* Cambridge, Mass.: Harvard University Press, 1999.

———. *The Mind Has No Sex? Women in the Origins of Modern Science.* Cambridge, Mass.: Harvard University Press, 1989.

———. *Nature's Body: Gender in the Making of Modern Science.* Boston: Beacon Press, 1993.

———. *Plants and Empire: Colonial Bioprospecting in the Atlantic World.* Cambridge, Mass.: Harvard University Press, 2004.

Schiebinger, Londa, Angela N. H. Creager, and Elizabeth Lunbeck, eds. *Feminism in Twentieth-Century Science, Technology, and Medicine.* Chicago: University of Chicago Press, 2001.

Schiebinger, Londa, and Claudia Swan, eds. *Colonial Botany: Science, Commerce, and Politics in the Early Modern World.* Philadelphia: University of Pennsylvania Press, 2004.

Schuster, John A., and Richard R. Yeo, eds. *The Politics and Rhetoric of Scientific Method: Historical Studies.* Dordrecht: Reidel, 1986.

Schwartz, Charles. "Political Structuring of the Institutions of Science." In *Naked Science*, ed. Laura Nader, 148–59. New York: Routledge, 1996.

Scott, Catherine V. *Gender and Development: Rethinking Modernization and Dependency Theory.* Boulder, Colo.: Lynne Rienner, 1995.

Scott, Colin. "Science for the West, Myth for the Rest?" In *Naked Science*, ed. Laura Nader. New York: Routledge, 1996.

Scott, Hilda. *Does Socialism Liberate Women?* Boston: Beacon, 1974.

Seager, Joni. *Earth Follies: Coming to Feminist Terms with the Global Environmental Crisis.* New York: Routledge, 1993.

———. "Rachel Carson Died of Breast Cancer: The Coming of Age of Feminist Environmentalism." *Signs: Journal of Women in Culture and Society* 28, no. 3. (2003): 945–72.

Selin, Helaine, ed. *Encyclopedia of the History of Science, Technology, and Medicine in Non-Western Cultures.* Dordrecht: Kluwer, 1997.

Sen, Gita, and Caren Grown. *Development Crises and Alternative Visions: Third World Women's Perspectives.* New York: Monthly Review Press, 1987.

Shapin, Steven. *A Social History of Truth.* Chicago: University of Chicago Press, 1994.

Shapin, Steven, and Simon Schaffer. *Leviathan and the Air Pump.* Princeton: Princeton University Press, 1985.

Shils, Edward, ed. *Center and Periphery: Essays in Macrosociology.* Chicago: University of Chicago Press, 1975.

Shinn, T., J. Spaapen, and Raoul Waast, eds. *Science and Technology in a Developing World.* Dordrecht: Kluwer, 1996.

Shiva, Vandana. *Close to Home: Women Reconnect Ecology, Health, and Development Worldwide.* Philadelphia: New Society, 1994.

———. *Monocultures of the Mind: Perspectives on Biodiversity and Biotechnology.* New York and Penang, Malaysia: Zed Books and Third World Network, 1993.

———. *Staying Alive: Women, Ecology and Development.* London: Zed Books, 1989.

———. *Stolen Harvest: The Hijacking of the Global Food Supply.* Cambridge, Mass.: South End Press, 2000.

Smith, Dorothy E. *The Conceptual Practices of Power: A Feminist Sociology of Knowledge.* Boston: Northeastern University Press, 1990.

———. *The Everyday World as Problematic: A Sociology for Women.* Boston: Northeastern University Press, 1987.

———. *Institutional Ethnography: A Sociology for People.* Lanham, Md.: Rowman and Littlefield, 2005.

———. *Texts, Facts, and Femininity: Exploring the Relations of Ruling.* New York: Routledge, 1990.

———, ed. *Writing the Social: Critique, Theory, and Investigations.* Toronto: University of Toronto Press, 1999.

Smith, Linda Tuhiwai. *Decolonizing Methodology: Research and Indigenous Peoples.* New York: St. Martin's Press, 1999.

Smith, Mike, et al. "The Reinvention of Politics: Ulrich Beck and Reflexive Modernity." *Environmental Politics* 8, no. 3 (1999): 169–73.

Snively, Gloria, and John Corsiglia. "Discovering Indigenous Science: Implications for Science Education." In "Multicultural Sciences," ed. Glen Aikenhead, special issue, *Journal of Science Education* 85 (2001): 6–34.

Snow, C. P. *The Two Cultures: And a Second Look.* Cambridge: Cambridge University Press, 1964 (1959).

Sokal, Alan, and Jean Bricmont. *Fashionable Nonsense: Postmodern Intellectuals' Abuse of Science.* New York: Picador USA, 1998.

Sparr, Pamela, ed. *Mortgaging Women's Lives: Feminist Critiques of Structural Adjustment.* London: Zed Books, 1994.

Spivak, Gayatri. "Can the Subaltern Speak?" In *Marxism and the Interpretation of Culture,* ed. Cary Nelson and Lawrence Grossberg. Urbana: University of Illinois Press, 1988.

Stanley, William, and Nancy Brickhouse. "Teaching Science: The Multicultural Question Revisited." *Science Education* 85, no. 1 (2001).

Steingraber, Sandra. *Living Downstream.* New York: Vintage, 1997.

Steinmetz, George, ed. *The Politics of Method in the Human Sciences: Positivism and Its Epistemological Others.* Durham, N.C.: Duke University Press, 2005.

Stepan, Nancy Leys. *The Idea of Race in Science: Great Britain, 1800–1960.* London: Macmillan, 1982.

———. "Race and Gender: The Role of Analogy in Science." *Isis* 77 (1986).

Strum, Shirley, and Linda Fedigan. *Primate Encounters.* Chicago: University of Chicago Press, 2000.

Subramaniam, Banu. *A Question of Variation: Race, Gender and the Practice of Science.* Forthcoming.

Tenner, Edward. *Why Things Bite Back: Technology and the Revenge Effect.* London: Fourth Estate, 1996.

Terrall, Mary. "Heroic Narratives of Quest and Discovery." *Configurations* 6, no. 2 (1998): 223–42.

Third World Network. *Modern Science in Crisis: A Third World Response.* Penang, Malaysia: Third World Network, 1988. (Reprinted in *The Racial Economy of Science,* ed. Sandra Harding. Bloomington: Indiana University Press, 1993.)

Tickner, J. Ann. *Gendering World Politics.* New York: Columbia University Press, 2001.

——. *Self-Reliance versus Power Politics: The American and Indian Experiences in Building Nation States.* New York: Columbia University Press, 1987.

Tobach, Ethel, and Betty Rosoff, eds. *Genes and Gender.* Vols. 1–4. New York: Gordian Press, 1978, 1979, 1981, 1984.

Traweek, Sharon. *Beamtimes and Life Times.* Cambridge, Mass.: MIT Press, 1988.

Tuana, Nancy, and Shannon Sullivan, eds. "Epistemologies and Ethics of Ignorance." Special issue, *Hypatia: A Journal of Feminist Philosophy* 21, no. 3 (2006).

Turnbull, David. *Masons, Tricksters, and Cartographers: Comparative Studies in the Sociology of Science and Indigenous Knowledge.* New York: Harwood Academic Publishers, 2000.

Van den Daele, W. "The Social Construction of Science." In *The Social Production of Scientific Knowledge*, ed. E. Mendelsohn, P. Weingart, and R. Whitley. Dordrecht: Reidel, 1977.

Verran, Helen. *Science and an African Logic.* Chicago: University of Chicago Press, 2001.

Visvanathan, Nalini, Lynn Duggan, Laurie Nisonoff, and Nan Wiegersma, eds. *The Women, Gender and Development Reader.* London: Zed Books, 1997.

Waast, Roland, ed. *Sciences in the South: Current Issues.* Paris: ORSTOM Editions, 1996.

Wagner, Peter. "Certainty and Order, Liberty and Contingency." In *The Rise of the Social Sciences and the Formation of Modernity*, ed. Johan Heilbron, Lars Magnusson, and Bjorn Wittrock. Dordrecht: Kluwer, 1998.

Wajcman, Judy. *Feminism Confronts Technology.* University Park: Penn State University Press, 1991.

——. "Reflections on Gender and Technology Studies: In What State Is the Art?" *Social Studies of Science* 30, no. 3 (2000): 447–64.

——. *TechnoFeminism.* Cambridge, U.K.: Polity Press, 2004.

Wakhungu, Judi Wangalwa, and Elizabeth Cecelski. "A Crisis in Power: Energy Planning for Development." In *Missing Links*, ed. Gender Working Group. Ottawa: International Development Research Centre, Intermediate Technology Publications, and UNIFEM, 1995.

Wallerstein, Immanuel. *The Modern World-System.* Vol. 1. New York: Academic Press, 1974.

Warshaw, Robin. *I Never Called It Rape.* New York: Harper Perennial, 1994.

Watson, James. *The Double Helix.* New York: New American Library, 1969.

Watson-Verran, Helen, and David Turnbull. "Science and Other Indigenous Knowledge Systems." In *Handbook of Science and Technology Studies*, ed. Sheila Jasanoff, G. Markle, T. Pinch, and J. Petersen, 115–39. Thousand Oaks, Calif.: Sage, 1995.

Weatherford, Jack McIver. *Indian Givers: What the Native Americans Gave to the World*. New York: Crown, 1988.

Weinberg, Steven. "The Revolution that Didn't Happen." *New York Review of Books*, 8 October 1998, 48–52.

Wellman, David. *Portraits of White Racism*. New York: Cambridge University Press, 1977.

Williams, Eric. *Capitalism and Slavery*. Chapel Hill: University of North Carolina Press, 1944.

Williams, Patrick, and Laura Chrisman. *Colonial Discourse and Post-Colonial Theory*. New York: Columbia University Press, 1994.

Willinsky, John. *Learning to Divide the World: Education at Empire's End*. Minneapolis: University of Minnesota Press, 1998.

Winner, Langdon. "Do Artefacts Have Politics?" *Daedalus* 109 (1980): 121–36.

Wittrock, Bjorn. "Early Modernities: Varieties and Transitions." *Daedalus* 127, no. 3 (1998): 19–40.

———. "Modernity: One, None, or Many? European Origins and Modernity as a Global Condition." In "Multiple Modernities," ed. S. N. Eisenstadt, special issue, *Daedalus* 129, no. 1 (2000): 31–60.

Wolf, Diane L. *Feminist Dilemmas in Fieldwork*. Boulder, Colo.: Westview Press, 1996.

Wolf, Eric. *Europe and the People without a History*. Berkeley: University of California Press, 1984.

Woolgar, Steve, ed. *Knowledge and Reflexivity*. Beverly Hills, Calif.: Sage, 1988.

———. *Science: The Very Idea*. New York: Tavistock, 1988.

Wyer, Mary, et al. *Women, Science, and Technology: A Reader in Feminist Science Studies*. New York: Routledge, 2001.

Wylie, Alison. "Why Standpoint Matters." In *Science and Other Cultures*, ed. Robert Figueroa and Sandra Harding. New York: Routledge, 2003.

Wylie, Philip. *Generation of Vipers*. New York: Holt, Rinehart and Winston, 1978 (1942, 1955).

Yoon, Soon-Young. "Looking at Health Through Women's Eyes." In *Missing Links*, ed. Gender Working Group. Ottawa: International Development Research Centre, Intermediate Technology Publications, and UNIFEM, 1995.

INDEX

— — — —

American Academy of Arts and Sciences, 8–9
Amin, Samir, 148

Beck, Ulrich, 17, 24, 49–74, 218–19
Beck-Gersheim, Elisabeth, 26–27, 252 n. 4

Castells, Manuel, 244 n. 20
Caulfield, Mina Davis, 163, 198
"Centres of calculation," 24, 42
Cold War, end of, 25, 75
Colonial rule, end of, 15, 155, 253 n. 13
"Craft labor," 1, 187, 207

Davis, Angela, 163, 198
Delinking, 148–49
Dependency theory, 135, 203, 222
Differentiation: decrease in, 50, 86, 188–89, 219–20, 229; of social institutions, 177–78
Disaggregation. *See* Differentiation
Discrimination: cultural differences and, 14; against women in sciences, 103–7
Diversity, cognitive, 150

Engels, Friedrich, 14
Environmental knowledge, traditional (TEK), 16, 138–43, 160–62
Ethnoscience, comparative, 138–42
Exceptionalism, 3–5
Experience, everyday, 60–62

Fact/value distinction, 28–34, 47
Felski, Rita, 191–213, 253 n. 10
Freud, Sigmund, 210–11, 253 n. 13

Gender issues, 110–14, 117–22; science and, 7, 103, 106–7, 108–9, 117, 122–26
Gibbons, Michael, 17–18, 75–97, 219–20
Gilligan, Carol, 116
Green Movement, German, 17–18, 24, 65

Haraway, Donna, 26, 243 n. 18
Hartmann, Heidi, 255 n. 10
Hartsock, Nancy, 115, 200

Hobson, John, 132, 141–42
hooks, bell, 118
Households: distributed, 231–32; women's lives in, 225–34

Identity, European, 141–42
Identity politics, Latour and, 26, 37–40
Indigenous knowledge (ɪ ᴋ), 16, 138–43, 160–62; within modern societies, 200, 252 n. 5
Inkeles, Alex, 175, 208

Jameson, Fredric, 119
Jardine, Alice, 191–213, 253 n. 10

Kelly-Gadol, Joan, 193–99, 252 n. 3, 254 n. 2
Kochhar, R. K., 137–38
Kuhn, Thomas, 141

Latour, Bruno, 17–18, 23–48, 186, 245 n. 25
Lerner, Daniel, 175

MacKinnon, Catharine, 116, 119
Marx, Karl, 14
Marxian critique, 65–67
Merian, Maria Sibylla, 164, 168
Modernity (modernities): amodernity, 30, 217; counter-modernity, 63–65, 74; end of, 50, 86, 189; first, 49–74, 223; literary and cultural theory and, 11; modern philosophy and, 9; multiple, 9, 66, 68, 173–90; nonmodernity, 30; pre-modernity, 3, 8, 174; scientific revolution and, 9; second, 49–74; substantive, 11–12, 177–78; temporal, 8–11, 177–78
Mohanty, Chandra Talpade, 205
Murata, Junichi, 182–83, 206–7, 251 n. 8

Narayan, Uma, 127
Networks, hybrid, 29, 30, 45
North: development policies of, 164–67; scientific tradition of, 19, 103, 146–53
Nowotny, Helga, 17–18, 75–97, 219–220

Objectivity: associated with masculinity, 200–201; reflexivity and, 57–58, 135; standards of, 7; standpoint and, 144–45; strong, 114–22
"Oriental West." *See* Western civilization, Eastern origins of
Ortiz, Renato, 150

Parsons, Talcott, 175
Pateman, Carole, 195
Periodization, 194–96
Philip, Kavita, 164
Plato, myth of the cave, 32, 33, 47
Postmodernism, 2–3, 6, 11, 12, 18, 25, 30, 47, 126, 218, 223
Prakash, Gyan, 150
Public and private spheres, 11, 106, 126, 204–5

Quine, Willard Van Orman, 141, 241 n. 3

Rationality: expertise and, 209; pluralism of, 12; scientific, 7, 73, 204–5
Reflexivity, 49–74, 255 n. 6; robust, 124–25
Risk society, 49–51, 70; everyday experience and, 60–61; modernity and, 57, 62–63; modern sciences and, 53–55
Rodney, Walter, 247–48 n. 10
Rossiter, Margaret, 168
Rostow, W. W., 175, 208
Rouse, Joseph, 206–7, 245 n. 23
Ruddick, Sara, 250 n. 6

Salomon, Jean-Jacques, 77
Schiebinger, Londa, 161, 164, 168

Science(s), 16; education, 108; Mode 1, 78, 80–85, 90–96; Mode 2, 76–85, 88, 94; multiple, 145–46; North and South traditions of, 146–51; "science of questions," 70–71; "science of science," 58–60, 144; Science Wars, 6, 152–54
Scott, Catherine V., 195, 205, 210, 253 n. 10
Scott, Peter, 17–18, 75–97, 219–20
Shils, Edward, 175
Social constructivism: multiculturalism and, 31; political engagement and, 245 n. 23; science studies and, 47, 107–8; Science Wars and, 152–53; technology studies and, 6, 179–83
Social sciences, 10; modernization and, 192, 253 n. 13
Standpoint theory, 14, 109–10, 114–22, 224

Steingraber, Sandra, 219
"Suturing," 1, 179–87

Traditional environmental knowledge. *See* Environmental knowledge, traditional
Traweek, Sharon, 251 n. 10
Triumphalism, 3–5, 151, 154, 174, 176, 177, 214, 221, 224, 233

Voyages of discovery, 42, 56, 131–32, 136, 162–64

Wallerstein, Immanuel, 203
Western civilization, Eastern origins of, 1, 131–32, 141–42, 176, 188, 251 n. 2
Williams, Eric, 136
Wittrock, Bjorn, 9
World systems theory, 135–36, 203, 208
Wylie, Philip, 253–254 n. 13

SANDRA HARDING is a professor of women's studies and education at the University of California, Los Angeles. Her many books include *Science and Social Inequality: Feminist and Postcolonial Issues*; *The Feminist Standpoint Theory Reader: Intellectual and Political Controversies*; and *Is Science Multicultural? Postcolonialisms, Feminisms, Epistemologies.*

Library of Congress Cataloging-in-Publication Data

Harding, Sandra G.
Sciences from below : feminisms, postcolonialities, and modernities / Sandra Harding.
p. cm. — (Next wave: New directions in women's studies)
Includes bibliographical references and index.
ISBN 978-0-8223-4259-5 (cloth : alk. paper)
ISBN 978-0-8223-4282-3 (pbk. : alk. paper)
1. Science—Social aspects. 2. Science—Philosophy.
3. Technology—Philosophy. 4. Philosophy—History.
5. Civilization, Modern. 6. Women in science.
7. Feminist theory. I. Title.
Q175.5.H395 2008
306.4'5—dc22 2008003008